TEACHING ABOUT

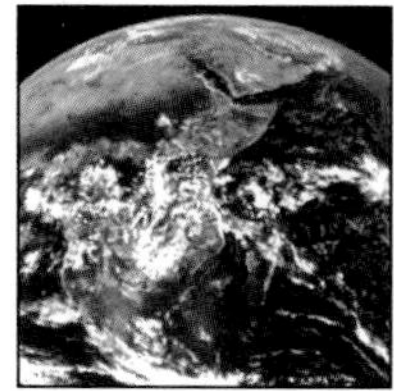

EVOLUTION

AND THE

NATURE OF

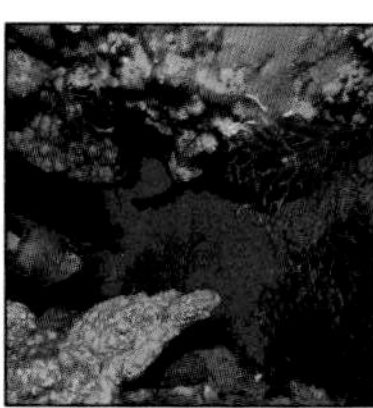

SCIENCE

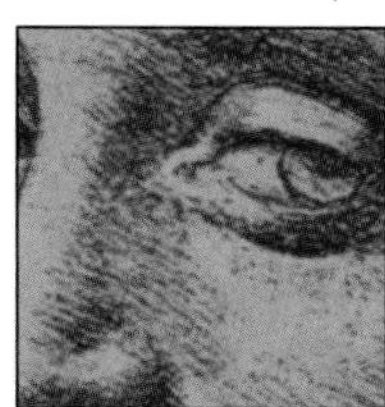

NATIONAL ACADEMY PRESS

Washington, DC

NATIONAL ACADEMY PRESS 2101 Constitution Avenue, NW Washington, DC 20418

The National Academy of Sciences is a private, nonprofit, self-perpetuating society of distinguished scholars engaged in scientific and engineering research, dedicated to the furtherance of science and technology and to their use for the general welfare. Upon the authority of the charter granted to it by the Congress in 1863, the Academy has a mandate that requires it to advise the federal government on scientific and technical matters.

Library of Congress Cataloging-in-Publication Data

Teaching about evolution and the nature of science / [Working Group on Teaching Evolution, National Academy of Sciences].

p. cm.

Includes bibliographical references and index.

ISBN 0-309-06364-7 (pbk.)

1. Evolution (Biology)—Study and teaching. 2. Science—Study and teaching. I. National Academy of Sciences (U.S.). Working Group on Teaching Evolution.

QH362.T435 1998

576.8′071—dc21 98-16100

CIP

Printed in the United States of America

Teaching About Evolution and the Nature of Science is available for sale from the National Academy Press, 2101 Constitution Avenue, NW, Box 285, Washington, DC 20055.

Call 1-800-624-6242 or 202-334-3313 (in the Washington Metropolitan Area).

The report is also available online at www.nap.edu/readingroom/books/evolution98

Working Group on Teaching Evolution

Donald Kennedy *(Chairman)*
Bing Professor of Environmental Studies
Stanford University
Stanford, California

Bruce Alberts
President
National Academy of Sciences
Washington, DC

Danine Ezell
Science Department
Bell Junior High School
San Diego, California

Tim Goldsmith
Department of Biology
Yale University
New Haven, Connecticut

Robert Hazen
Staff Scientist, Geophysical Laboratory
Carnegie Institution of Washington
Washington, DC

Norman Lederman
Professor, College of Science
Science and Mathematics Education
Oregon State University
Corvallis, Oregon

Joseph McInerney
Director
Biological Sciences Curriculum Study
Colorado Springs, Colorado

John Moore
Professor Emeritus of Biology
University of California
Riverside, California

Eugenie Scott
Executive Director
National Center for Science Education
El Cerrito, California

Maxine Singer
President
Carnegie Institution of Washington
Washington, DC

Mike Smith
Associate Professor of Medical Education
Mercer University School of Medicine
Macon, Georgia

Marilyn Suiter
Director, Education and Human Resources
American Geological Institute
Alexandria, Virginia

Rachel Wood
Science Specialist
Delaware State Department of Public Instruction
Dover, Delaware

Staff of the Center for Science, Mathematics, and Engineering Education:

Rodger Bybee, Executive Director
Peggy Gill, Research Assistant
Jay Hackett, Visiting Fellow
Patrice Legro, Division Director
Steve Olson, Consultant Writer

THE NATIONAL ACADEMY OF SCIENCES
WASHINGTON, DC

Visit us at
www.nas.edu

Acknowledgments

The National Academy of Sciences gratefully acknowledges contributions from:

Howard Hughes Medical Institute

The Esther A. and Joseph Klingenstein Fund, Inc.

The Council of the National Academy of Sciences

The 1997 Annual Fund of the National Academy of Sciences,
whose donors include
NAS members and other science-interested individuals.

We also extend special thanks to members of the
Council of State Science Supervisors
and teachers who participated in focus groups and provided guidance
on the development of this document.

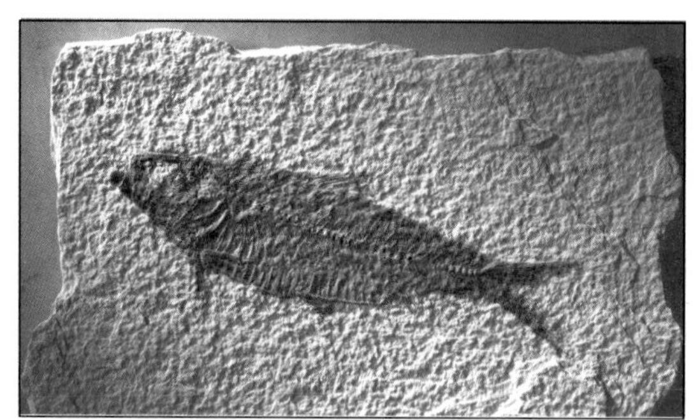

Contents

Preface

In a 1786 letter to a friend, Thomas Jefferson called for "the diffusion of knowledge among the people. No other sure foundation can be devised for the preservation of freedom and happiness."[1] Jefferson saw clearly what has become increasingly evident since then: the fortunes of a nation rest on the ability of its citizens to understand and use information about the world around them.

We are about to enter a century in which the United States will be even more dependent on science and technology than it has been in the past. Such a future demands a citizenry able to use many of the same skills that scientists use in their work—close observation, careful reasoning, and creative thinking based on what is known about the world.

The ability to use scientific knowledge and ways of thinking depends to a considerable extent on the education that people receive from kindergarten through high school. Yet the teaching of science in the nation's public schools often is marred by a serious omission. Many students receive little or no exposure to the most important concept in modern biology, a concept essential to understanding key aspects of living things—biological evolution. People and groups opposed to the teaching of evolution in the public schools have pressed teachers and administrators to present ideas that conflict with evolution or to teach evolution as a "theory, not a fact." They have persuaded some textbook publishers to downplay or eliminate treatments of evolution and have championed legislation and policies at the state and local levels meant to discourage the teaching of evolution.

These pressures have contributed to widespread misconceptions about the state of biological understanding and about what is and is not science. Fewer than one-half of American adults believe that humans evolved from earlier species.[2] More than one half of Americans say that they would like to have creationism taught in public school classrooms[3]—even though the Supreme Court has ruled that "creation science" is a religious idea and that its teaching cannot be mandated in the public schools.[4]

The widespread misunderstandings about evolution and the conviction that creationism should be taught in science classes are of great concern to the National Academy of Sciences, a private nonpartisan group of 1,800 scientists dedicated to the use of science and technology for the general welfare. The Academy and its affiliated institutions—the National Academy of Engineering, the Institute of Medicine, and the National Research Council—have all sought to counter misinformation about evolution because of the enormous body of data supporting evolution and because of the importance of evolution as a central concept in understanding our planet.

The document that you are about to read is addressed to several groups at the center of the ongoing debate over evolution: the teachers, other educators, and policy makers who design, deliver, and oversee classroom instruction in biology. It summarizes the overwhelming observational evidence for evolution and suggests effective ways of teaching the subject. It explains the nature of science and describes how science differs from other human endeavors. It provides answers to frequently asked questions about evolution and the nature of science and offers guidance on how to analyze and select teaching materials.

This publication does not attempt specifically to refute the ideas proffered by those who oppose the teaching of evolution in public schools. A related document, *Science and Creationism: A View from the National Academy of Sciences,* discusses evolution and "creation science" in detail.[5] This publication instead provides information and resources that teachers and administrators can use to inform themselves, their students, parents, and others about evolution and the role of science in human affairs.

One source of resistance to the teaching of evolution is the belief that evolution conflicts with religious principles. But accepting evolution as an accurate description of the history of life on earth does not mean rejecting religion. On the contrary, most religious communities do not hold that the concept of evolution is at odds with their descriptions of creation and human origins.

Nevertheless, religious faith and scientific knowledge, which are both useful and important, are different. This publication is designed to help ensure that students receive an education in the sciences that reflects this distinction.

The book is divided into seven chapters and five appendices, plus three interspersed "dialogues" in which several fictional teachers discuss the implications of the ideas discussed in the book.

• Chapter 1, "Why Teach Evolution," introduces the basic concepts of evolutionary theory and provides scientific definitions of several common terms, such as "theory" and "fact," used throughout the book.

• The first dialogue, "The Challenge to Teachers," follows the conversation of three teachers as they discuss some of the problems that can arise in teaching evolution and the nature of science.

• Chapter 2, "Major Themes in Evolution," provides a general overview of evolutionary processes, describes the evidence supporting evolution, and shows how evolutionary theory is related to other areas of biology.

• The second dialogue, "Teaching About the Nature of Science," follows the three teachers as they engage in a teaching exercise designed to demonstrate several prominent features of science.

• Chapter 3, "Evolution and the Nature of Science," uses several scientific theories, including evolution, to highlight important characteristics of scientific endeavors.

• The third dialogue, "Teaching Evolution Through Inquiry," presents a teacher using an exercise designed to interest and educate her students in fossils and the mechanisms of evolution.

• Chapter 4, "Evolution and the *National Science Education Standards*," begins by describing the recent efforts to specify what students should know and be able to do as a result of their education in the sciences. It then reproduces sections from the 1996 *National Science Education Standards* released by the National Research Council that relate to biological evolution and the nature and history of science.

• Chapter 5, "Frequently Asked Questions About Evolution and the Nature of Science," gives short answers to some of the questions asked most frequently by students, parents, educators, and others.

• Chapter 6, "Activities for Teaching About Evolution and the Nature of Science," provides eight sample activities that teachers can use to develop students' understanding of evolution and scientific inquiry.

• Chapter 7, "Selecting Instructional Materials," lays out criteria that can be used to evaluate school science programs and the content and design of instructional materials.

• The appendices summarize significant court decisions regarding evolution and creationism issues, reproduce statements from a number of organizations regarding the teaching of evolution, provide references for further reading and other resources, and conclude with a list of reviewers.

Teaching About Evolution and the Nature of Science was produced by the Working Group on Teaching Evolution under the Council of the National Academy of Sciences. The Working Group consists of 13 scientists and educators who have been extensively involved in research and education on evolution and related scientific subjects. The group worked closely with teachers, school administrators, state officials, and others in preparing this publication, soliciting suggestions for what would be most useful, and responding to comments on draft materials. We welcome additional input and guidance from readers that we can incorporate into future versions of this publication. Please visit our World Wide Web site at **www4.nas.edu/opus/evolve.nsf** for additional information.

Notes

1. Thomas Jefferson, To George Wythe, "Crusade Against Ignorance" in *Thomas Jefferson on Education*, ed. Gordon C. Lee. 1961. New York: Teachers College Press, pp. 99-100.
2. National Science Board. 1996. *Science and Engineering Indicators*—1996. Washington, DC: U.S. Government Printing Office.
3. Gallup Poll, News Release, May 24, 1996.
4. In the 1987 case *Edwards v. Aguillard*, the U.S. Supreme Court reaffirmed the 1982 decision of a federal district court that the teaching of "creation science" in public schools violates the First Amendment of the U.S. Constitution.
5. National Academy of Sciences. (in press). *Science and Creationism: A View from the National Academy of Sciences.* Washington, DC: National Academy Press. (See www.nap.edu)

1

Why Teach Evolution?

Why is it so important to teach evolution? After all, many questions in biology can be answered without mentioning evolution: How do birds fly? How can certain plants grow in the desert? Why do children resemble their parents? Each of these questions has an immediate answer involving aerodynamics, the storage and use of water by plants, or the mechanisms of heredity. Students ask about such things all the time.

The answers to these questions often raise deeper questions that are sometimes asked by students: How did things come to be that way? What is the advantage to birds of flying? How did desert plants come to differ from others? How did an individual organism come to have its particular genetic endowment? Answering questions like these requires a historical context—a framework of understanding that recognizes change through time.

People who study nature closely have always asked these kinds of questions. Over time, two observations have proved to be especially perplexing. The older of these has to do with the diversity of life: Why are there so many different kinds of plants and animals? The more we explore the world, the more impressed we are with the multiplicity of kinds of organisms. In the mid-nineteenth century, when Charles Darwin was writing *On the Origin of Species*, naturalists recognized several tens of thousands of different plant and animal species. By the middle of the twentieth century, biologists had paid more attention to less conspicuous forms of life, from insects to microorganisms, and the estimate was up to 1 or 2 million. Since then, investigations in tropical rain forests—the center of much of the world's biological diversity—have multiplied those estimates at least tenfold. What process has created this extraordinary variety of life?

The second question involves the inverse of life's diversity. How can the similarities among organisms be explained? Humans have always noticed the similarities among closely related species, but it gradually became apparent that even distantly related species share many anatomical and functional characteristics. The bones in a whale's front flippers are arranged in much the same way as the bones in our own arms. As organisms grow from fertilized egg cells into embryos, they pass through many similar developmental stages. Furthermore, as paleontologists studied the fossil record, they discovered countless extinct species that are clearly related in various ways to organisms living today.

This question has emerged with even greater force as modern experimental biology has focused on processes at the cellular and molecular level. From bacteria to yeast to mice to humans, all living things use the same biochemical machinery to carry out the basic processes of life. Many of the proteins that make up cells and catalyze chemical reactions in the body are virtually identical across species. Certain human genes that code for proteins differ little from the corresponding genes in fruit flies,

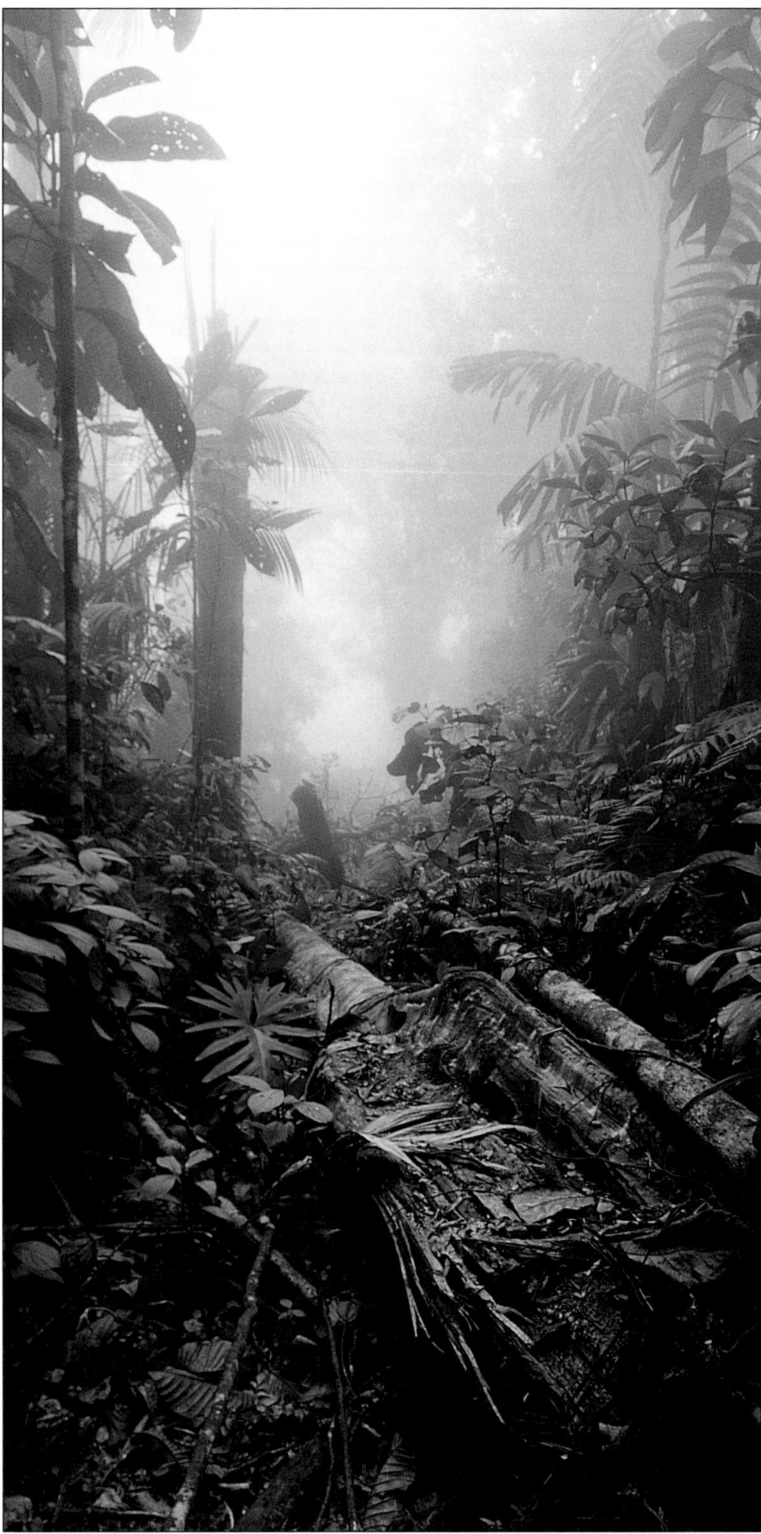

Investigations of forest ecosystems have helped reveal the incredible diversity of earth's living things.

mice, and primates. All living things use the same biochemical system to pass genetic information from one generation to another.

From a scientific standpoint, there is one compelling answer to questions about life's commonalities. Different kinds of organisms share so many characteristics of structure and function because they are related to one another. But how?

Solving the Puzzle

The concept of biological evolution addresses both of these fundamental questions. It accounts for the relatedness among organisms by explaining that the millions of different species of plants, animals, and microorganisms that live on earth today are related by descent from common ancestors—like distant cousins. Organisms in nature typically produce more offspring than can survive and reproduce given the constraints of food, space, and other resources in the environment. These offspring often differ from one another in ways that are heritable—that is, they can pass on the differences genetically to their own offspring. If competing offspring have traits that are advantageous in a given environment, they will survive and pass on those traits. As differences continue to accumulate over generations, populations of organisms diverge from their ancestors.

This straightforward process, which is a natural consequence of biologically reproducing organisms competing for limited resources, is responsible for one of the most magnificent chronicles known to science. Over billions of years, it has led the earliest organisms on earth to diversify into all of the plants, animals, and microorganisms that exist today. Though humans, fish, and bacteria would seem to be so different as to defy comparison, they all share some of the characteristics of their common ancestors.

Evolution also explains the great diversity of modern species. Populations of organisms

Living fish and fossil fish share many similarities, but the fossil fish clearly belongs to a different species that no longer exists. The progression of species found in the fossil record provides powerful evidence for evolution.

with characteristics enabling them to occupy ecological niches not occupied by similar organisms have a greater chance of surviving. Over time—as the next chapter discusses in more detail—species have diversified and have occupied more and more ecological niches to take advantage of new resources.

Evolution explains something else as well. During the billions of years that life has been on earth, it has played an increasingly important role in altering the planet's physical environment. For example, the composition of our atmosphere is partly a consequence of living systems. During photosynthesis, which is a product of evolution, green plants absorb carbon dioxide and water, produce organic compounds, and release oxygen. This process has created and continues to maintain an atmosphere rich in oxygen. Living communities also profoundly affect weather and the movement of water among the oceans, atmosphere, and land. Much of the rainfall in the forests of the western Amazon basin consists of water that has already made one or more recent trips through a living plant. In addition, plants and soil microorganisms exert important controls over global temperature by absorbing or emitting "greenhouse gases" (such as carbon dioxide and methane) that increase the earth's capacity to retain heat.

In short, biological evolution accounts for three of the most fundamental features of the world around us: the similarities among living things, the diversity of life, and many features of the physical world we inhabit. Explanations of these phenomena in terms of evolution draw on results from physics, chemistry, geology, many areas of biology, and other sciences. Thus, evolution is the central organizing principle that biologists use to understand the world. To teach biology without explaining evolution deprives students of a powerful concept that brings great order and coherence to our understanding of life.

The teaching of evolution also has great practical value for students. Directly or indirectly, evolutionary biology has made many contributions to society. Evolution explains why many human pathogens have been developing resistance to formerly effective drugs and suggests ways of confronting this increasingly serious problem (this issue is discussed in greater detail in Chapter 2). Evolutionary biology has also

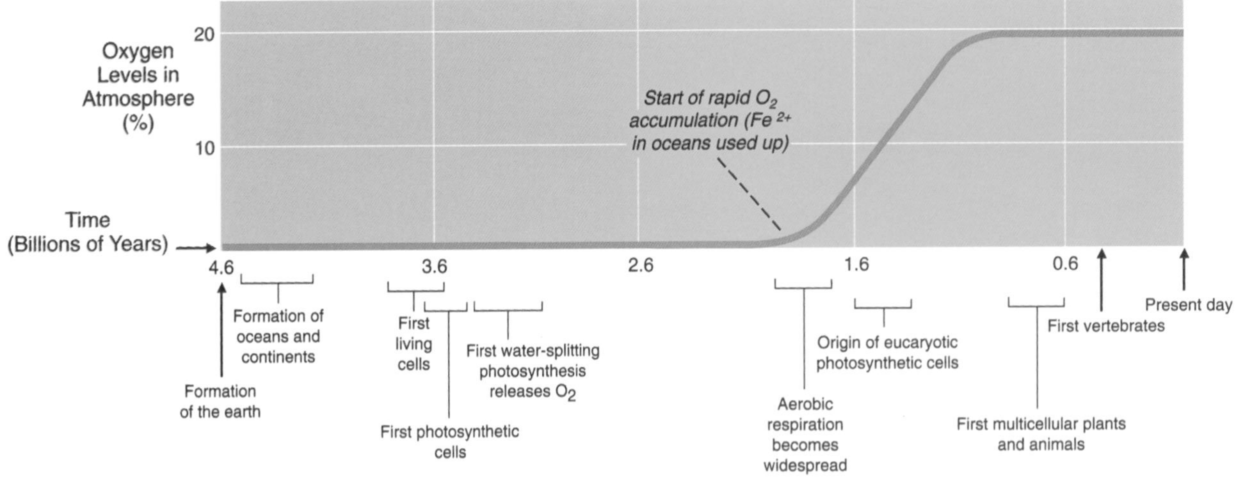

Living things have altered the earth's oceans, land surfaces, and atmosphere. For example, photosynthetic organisms are responsible for the oxygen that makes up about a fifth of the earth's atmosphere. The rapid accumulation of atmospheric oxygen about 2 billion years ago led to the evolution of more structured eucaryotic cells, which in turn gave rise to multicellular plants and animals.

contributed to many important agricultural advances by explaining the relationships among wild and domesticated plants and animals and their natural enemies. An understanding of evolution has been essential in finding and using natural resources, such as fossil fuels, and it will be indispensable as human societies strive to establish sustainable relationships with the natural environment.

Such examples can be multiplied many times. Evolutionary research is one of the most active fields of biology today, and discoveries with important practical applications occur on a regular basis.

Those who oppose the teaching of evolution in public schools sometimes ask that teachers present "the evidence against evolution." However, there is no debate within the scientific community over whether evolution occurred, and there is no evidence that evolution has not occurred. Some of the details of how evolution occurs are still being investigated. But scientists continue to debate only the particular mechanisms that result in evolution, not the overall accuracy of evolution as the explanation of life's history.

Evolution and the Nature of Science

Teaching about evolution has another important function. Because some people see evolution as conflicting with widely held beliefs, the teaching of evolution offers educators a superb opportunity to illuminate the nature of science and to differentiate science from other forms of human endeavor and understanding.

Chapter 3 describes the nature of science in detail. However, it is important from the outset to understand how the meanings of certain key words in science differ from the way that those words are used in everyday life.

Think, for example, of how people usually use the word "theory." Someone might refer to an idea and then add, "But that's only a theory." Or someone might preface a remark by saying, "My theory is . . . ." In common usage, theory often means "guess" or "hunch."

In science, the word "theory" means something quite different. It refers to an overarching explanation that has been well substantiated. Science has many other powerful theories besides evolution. Cell theory says that all living things are composed of cells. The heliocentric theory says that the earth revolves around the sun rather than vice versa. Such concepts are supported by such abundant observational and experimental evidence that they are no longer questioned in science.

Sometimes scientists themselves use the word "theory" loosely and apply it to tentative explanations that lack well-established evidence. But it is important to distinguish these casual uses of the word "theory" with its use to describe concepts such as evolution that are supported by overwhelming evidence. Scientists might wish that they had a word other than "theory" to apply to such enduring explanations of the natural world, but the term is too deeply engrained in science to be discarded.

As with all scientific knowledge, a theory can be refined or even replaced by an alternative theory in light of new and compelling evidence. For example, Chapter 3 describes how the geocentric theory that the sun revolves around the earth was replaced by the heliocentric theory of the earth's rotation on its axis and revolution around the sun. However, ideas are not referred to as "theories" in science unless they are supported by bodies of evidence that make their subsequent abandonment very unlikely. When a theory is supported by as much evidence as evolution, it is held with a very high degree of confidence.

In science, the word "hypothesis" conveys the tentativeness inherent in the common use of the word "theory." A hypothesis is a testable statement about the natural world. Through experiment and observation, hypotheses can be supported or rejected. As the earliest level of understanding, hypotheses can be used to construct more complex inferences and explanations.

Like "theory," the word "fact" has a different meaning in science than it does in common usage. A scientific fact is an observation that has been confirmed over and over. However, observations are gathered by our senses, which can never be trusted entirely. Observations also can change with better technologies or with better ways of looking at data. For example, it was held as a scientific fact for many years that human cells have 24 pairs of chromosomes, until improved techniques of microscopy revealed that they actually have 23. Ironically, facts in science often are more susceptible to change than theories—which is one reason why the word "fact" is not much used in science.

Finally, "laws" in science are typically descriptions of how the physical world behaves under certain circumstances. For example, the laws of motion describe how objects move when subjected to certain forces. These laws can be very useful in supporting hypotheses and theories, but like all elements of science they can be altered with new information and observations.

Glossary of Terms Used in Teaching About the Nature of Science

Fact: In science, an observation that has been repeatedly confirmed.

Law: A descriptive generalization about how some aspect of the natural world behaves under stated circumstances.

Hypothesis: A testable statement about the natural world that can be used to build more complex inferences and explanations.

Theory: In science, a well-substantiated explanation of some aspect of the natural world that can incorporate facts, laws, inferences, and tested hypotheses.

Evolution and Everyday Life

The concept of evolution has an importance in education that goes beyond its power as a scientific explanation. All of us live in a world where the pace of change is accelerating. Today's children will face more new experiences and different conditions than their parents or teachers have had to face in their lives.

The story of evolution is one chapter—perhaps the most important one—in a scientific revolution that has occupied much of the past four centuries. The central feature of this revolution has been the abandonment of one notion about stability after another: that the earth was the center of the universe, that the world's living things are unchangeable, that the continents of the earth are held rigidly in place, and so on. Fluidity and change have become central to our understanding of the world around us. To accept the probability of change—and to see change as an agent of opportunity rather than as a threat—is a silent message and challenge in the lesson of evolution.

The following dialogue dramatizes some of the problems educators encounter in teaching evolution and demonstrates ways of overcoming these obstacles. Chapter 2 returns to the basic themes that characterize evolutionary theory, and Chapter 3 takes a closer look at the nature of science.

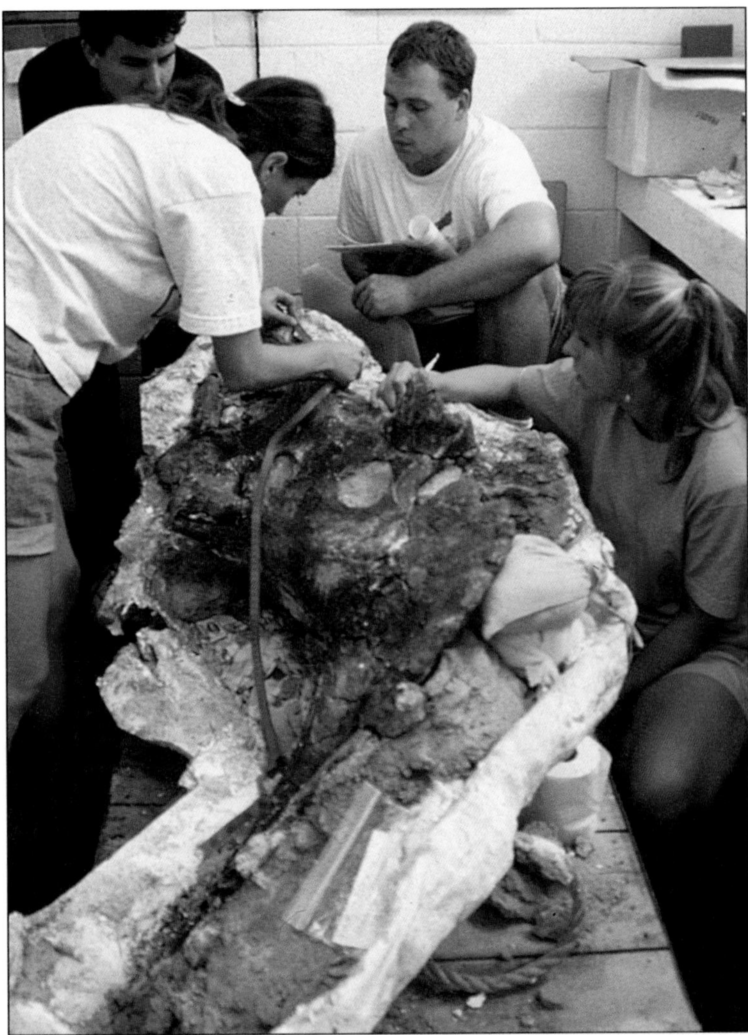

Scientists examining the head of *Chasmosaurus mariscalensis* hone their understanding of nature by comparing it against observations of the world. Clockwise from upper right: Prof. Paul Sereno, Univ. of Chicago; assistant Cathy Forster, Univ. of Chicago; students Hilary Tindle and Tom Evans, who discovered the skull in the field in March 1991 in Big Bend National Park, Texas.

Those who oppose the teaching of evolution often say that evolution should be taught as a "theory, not as a fact." This statement confuses the common use of these words with the scientific use. In science, theories do not turn into facts through the accumulation of evidence. Rather, theories are the end points of science. They are understandings that develop from extensive observation, experimentation, and creative reflection. They incorporate a large body of scientific facts, laws, tested hypotheses, and logical inferences. In this sense, evolution is one of the strongest and most useful scientific theories we have.

Dialogue

THE CHALLENGE TO TEACHERS

Teaching evolution presents special challenges to science teachers. Sources of support upon which teachers can draw include high-quality curricula, adequate preparation, exposure to information useful in documenting the evidence for evolution, and resources and contacts provided by professional associations.

One important source of support for teachers is to share problems and explore solutions with other teachers. The following vignette illustrates how a group of teachers—in this case, three biology teachers at a large public high school—can work together to solve problems and learn from each other.

••••••••••••••••••••••

It is the first week of classes at Central High School. As the bell rings for third period, Karen, the newest teacher on the faculty, walks into the teachers' lounge. She greets her colleagues, Barbara and Doug.

"How are your first few days going?" asks Doug.

"Fine," Karen replies. "The second-period Biology I class is full, but it'll be okay. By the way, Barbara, thanks for letting me see your syllabus for Bio I. But I wanted to ask you about teaching evolution—I didn't see it there."

"You didn't see it on my syllabus because it's not a separate topic," Barbara says. "I use evolution as a theme to tie the course together, so it comes into just about every unit. You'll see a section called 'History of Life' on the second page, and there's a section called 'Natural Selection.' But I don't treat evolution separately because it is related to almost every other topic in biology."[1]

"Wait a minute, Barbara," Doug says. "Is that good advice for a new teacher? I mean, evolution is a controversial subject, and a lot of us just don't get around to teaching it. I don't. You do, but you're braver than most of us."

"It's not a matter of bravery, Doug," Barbara replies. "It's a matter of what needs to be taught if we want students to understand biology. Teaching biology without evolution would be like teaching civics and never mentioning the United States Constitution."

"But how can you be sure that evolution is all that important. Aren't there a lot of scientists who don't believe in evolution? Say it's too improbable?"

"The debate in science is over some of the details of how evolution occurred, not whether evolution happened or not. A lot of science and science education organizations have made statements about why it is important to teach evolution...."[2]

"I saw a news report when I was a student," Karen interjects, "about a school district or state that put a disclaimer against evolution in all their biology textbooks. It said that students didn't need to believe in evolution because it wasn't a fact, only a theory. The argument was that no one really knows how life began or how it evolved because no one was there to see it happen."[3]

"If I taught evolution, I'd sure teach it as a theory—not a fact," says Doug.

"Just like gravity," Barbara says.

"Now, Barbara, gravity is a fact, not a theory."

"Not in scientific terms. The fact is that things fall. The explanation for why things fall is the theory of gravitation. Our problem is definitions. You're using 'fact' and 'theory' the way we use them in everyday life, but we need to use them as scientists use them. In science, a 'fact' is an observation that has

A fossil of *Archaeopteryx*, a bird that lived about 150 million years ago and had many reptilian characteristics, was discovered in 1861 and helped support the hypothesis of evolution proposed by Charles Darwin in *The Origin of Species* two years earlier.

been made so many times that it's assumed to be okay. How facts are explained is where theories come in: theories are explanations of what we observe. One place where students get confused about evolution is that they think of 'theory' as meaning 'guess' or 'hunch.' But evolution isn't a hunch. It's a scientific explanation, and a very good one."

"But how good a theory is it?" asks Doug. "We don't know everything about evolution."

"That's true," says Karen. "A student in one of my classes at the university told me that there are big gaps in the fossil record. Do you know anything about that?"

"Well, there's *Archaeopteryx*," says Doug. "It's a fossil that has feathers like a bird but the skeleton of a small dinosaur. It's one of those missing links that's not missing any more."

"In fact, there are good transitional fossils between primitive fish and amphibians and between reptiles and mammals," Barbara says. "Our knowledge of fossil intermediates is actually pretty good.[4] And, Doug, it sounds like you know more about evolution than you're letting on. Why don't you teach it?"

"I don't want any trouble. Every time I teach evolution, I have a student announce that 'evolution is against his religion.'"

"But most of the major religious denominations have taken official positions that accept evolution," says Barbara. "One semester a friend of mine in the middle school started out her Life Science unit by having her students interview their ministers or priests or rabbis about their religion's views on evolution. She said that most of her students came back really surprised. 'Hey,' they said, 'evolution is okay.' It defused the controversy in her class."

"She didn't have Stanley in her class," says Doug.

"Who's Stanley?" asks Karen.

"The son of a school board member. Given his family's religious views, I'm sure he would not come back saying evolution was okay."

"That can be a hard situation," says Barbara. "But even if Stanley came back to class saying that his religion does not accept evolution, it could help a teacher show that there are many different religious views about evolution. That's the point: religious people can still accept evolution."

"Stanley will never believe in evolution."

"We talk about 'believing' in evolution, but that's not necessarily the right word. We accept evolution as the best scientific explanation for a lot of observations—about fossils and biochemistry and evolutionary changes we can actually see, like how bacteria become resistant to certain medicines. That's why people accepted the idea that the earth goes around the sun—because it accounted for many different observations that we make. In science, when a better explanation comes around, it replaces earlier ones."

"Does that mean that evolution will be replaced by a better theory some day?" asks Karen.

"It's not likely. Not all old theories are

replaced, and evolution has been tested and has a lot of evidence to support it. The point is that doing science requires being willing to refine our theories to be consistent with new information."

"But there's still Stanley," says Doug. "He doesn't even want to hear about evolution."

"I had Stanley's sister in AP biology one year," Barbara replies. "She raised a fuss about evolution, and I told her that I wasn't going to grade her on her opinion of evolution but on her knowledge of the facts and concepts. She seemed satisfied with that and actually got an A in the class."

"I still think that if you teach evolution, it's only fair to teach both."

"What do you mean by both?" asks Barbara. "If you mean both evolution and creationism, what kind of creationism do you want to teach? Will you teach evolution and the Bible? What about other religions like Buddhism or the views of Native Americans? It's hard to argue for 'both' when there are a whole lot more than two options."

"I can't teach a whole bunch of creation stories in my Bio class," says Doug.

"That's the point. We can't add subjects to the science curriculum to be fair to groups that hold certain beliefs. Teaching ecology isn't fair to the polluter, either. Biology is a science class, and what should be taught is science."

"But isn't there something called 'creation science'?" asks Karen. "Can creationism be made scientific?"

"That's an interesting story. 'Creation science' is the idea that scientific evidence can support a literal interpretation of Genesis—that the whole universe was created all at once about 10,000 years ago."

"It doesn't sound very likely."

"It's not. Scientists have looked at the arguments and have found they are not supported by verifiable data. Still, back in the early 1980s, some states passed laws requiring that 'creation science' be taught whenever evolution was taught. But the Supreme Court threw out 'equal time' laws, saying that because creationism was inherently a religious and not a scientific idea, it couldn't be presented as 'truth' in science classes in the public schools."[5]

"Well, I'm willing to teach evolution," says Karen, "and I'd like to try it your way, Barbara, as a theme that ties biology together. But I really don't know enough about evolution to do it. Do you have any suggestions about where I can get information?"

"Sure, I'd be glad to share what I have. But an important part of teaching evolution has to do with explaining the nature of science. I'm trying out a demonstration after school today that I'm going to use with my Bio I class tomorrow. Why don't you both come by and we can try it out?"

"Okay," say Karen and Doug. "We'll see you then."

•••••••••••••••••••••

Barbara, Doug, and Karen's discussion of evolution and the nature of science resumes following Chapter 2.

Notes

1. The *National Science Education Standards* cite "evolution and equilibrium" as one of five central concepts that unify all of the sciences. (See www.nap.edu/readingroom/books/nses)
2. Appendix C contains statements from science and science education organizations that support the need to teach evolution.
3. In 1995, the Alabama board of education ordered that all biology textbooks in public schools carry inserts that read, in part, as follows: "This textbook discusses evolution, a controversial theory some scientists present as a scientific explanation for the origin of living things, such as plants, animals, and humans. No one was present when life first appeared on earth. Therefore, any statement about life's origins should be considered theory, not fact." Other districts have required similar disclaimers.
4. The book *From So Simple a Beginning: The Book of Evolution* by Philip Whitfield (New York: Macmillan, 1993) presents a well-illustrated overview of evolutionary history. *Evolution* by Monroe W. Strickberger (Boston: Jones and Bartlett, 2nd edition, 1995) is a thorough text written at the undergraduate level.
5. In the 1987 case *Edwards v. Aguillard*, the U.S. Supreme Court reaffirmed the 1982 decision of a federal district court that the teaching of "creation science" in public schools violates the First Amendment of the U.S. Constitution.

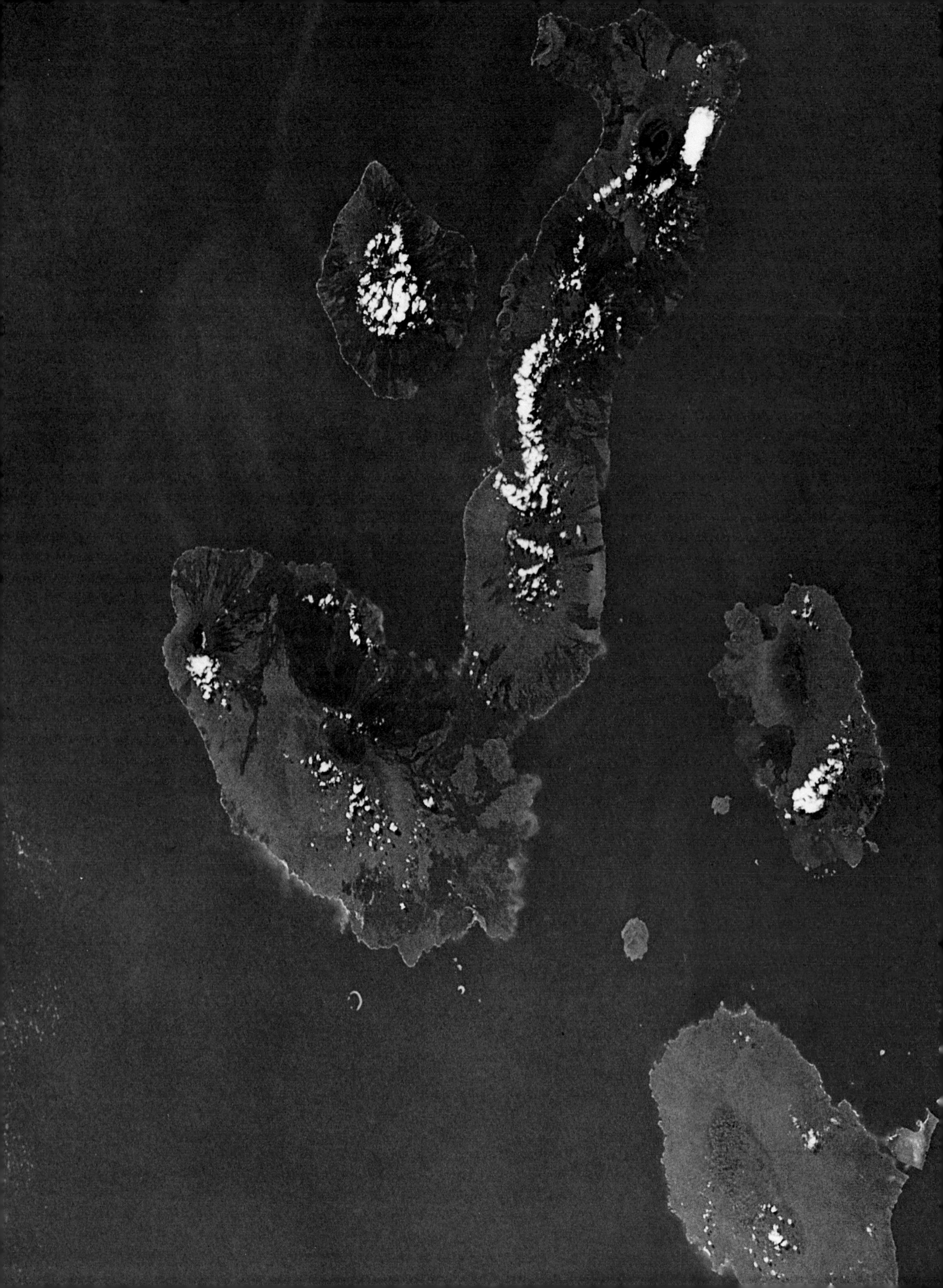

2

Major Themes in Evolution

The world around us changes. This simple fact is obvious everywhere we look. Streams wash dirt and stones from higher places to lower places. Untended gardens fill with weeds.

Other changes are more gradual but much more dramatic when viewed over long time scales. Powerful telescopes reveal new stars coalescing from galactic dust, just as our sun did more than 4.5 billion years ago. The earth itself formed shortly thereafter, when rock, dust, and gas circling the sun condensed into the planets of our solar system. Fossils of primitive microorganisms show that life had emerged on earth by about 3.8 billion years ago.

Similarly, the fossil record reveals profound changes in the kinds of living things that have inhabited our planet over its long history. Trilobites that populated the seas hundreds of millions of years ago no longer crawl about. Mammals now live in a world that was once dominated by reptilian giants such as *Tyrannosaurus rex*. More than 99 percent of the species that have ever lived on the earth are now extinct, either because all of the members of the species died, the species evolved into a new species, or it split into two or more new species.

Many kinds of cumulative change through time have been described by the term "evolution," and the term is used in astronomy, geology, biology, anthropology, and other sciences. This document focuses on the changes in living things during the long history of life on earth—on what is called biological evolution. The ancient Greeks were already speculating about the origins of life and changes in species over time. More than 2,500 years ago, the Greek philosopher Anaximander thought that a gradual evolution had created the world's organic coherence from a formless condition, and he had a fairly modern view of the transformation of aquatic species into terrestrial ones. Following the rise of Christianity, Westerners generally accepted the explanation provided in Genesis, the first book of the Judeo-Christian-Muslim Bible, that God created everything in its present form over the course of six days. However, other explanations existed even then. Among Christian theologians, for example, Saint Thomas Aquinas (1225 to 1274) stated that the earth had received the power to produce organisms and criticized the idea that species had originated in accordance with the timetables in Genesis.[1]

During the early 1800s, many naturalists speculated about changes in organisms, especially as geological investigations revealed the rich story laid out in the fossilized remains of extinct creatures. But although ideas about evolution were proposed, they never gained wide acceptance because no one was able to propose a plausible mechanism for how the form of an organism might change from one generation to another. Then, in 1858, two English naturalists—Charles Darwin and Alfred Russel Wallace—simultaneously issued papers proposing such a mechanism. Both

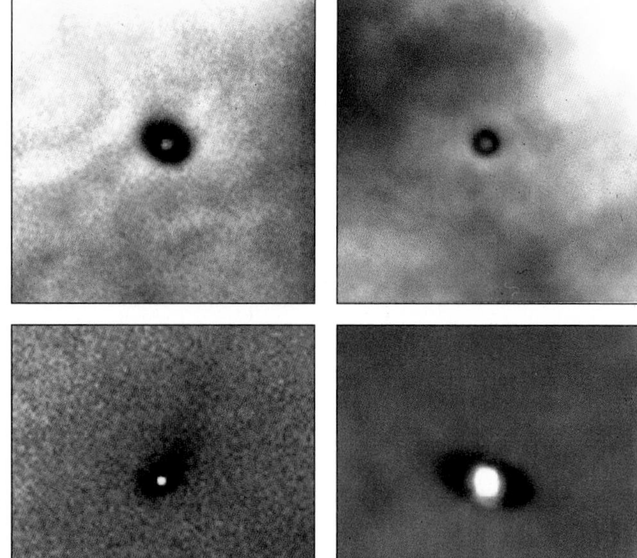

The Hubble Space Telescope has revealed many astronomical phenomena that ground-based telescopes cannot see. The images at right show disks of matter around young stars that could give rise to planets. In the image below, stars are forming in the tendrils of gas and dust extending from a gigantic nebula.

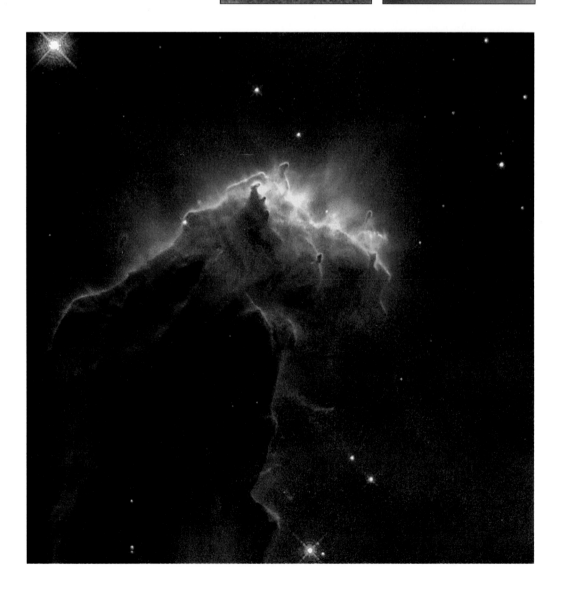

men observed that the individual members of a particular species are not identical but can differ in many ways. For example, some will be able to run a little faster, have a different color, or respond to the same circumstance in different ways. (Humans—including any class of high school students—have many such differences.) Both men further observed that many of these differences are inherited and can be passed on to offspring. This conclusion was evident from the experiences of plant and animal breeders.

Darwin and Wallace were both deeply influenced by the realization that, even though most species produce an abundance of offspring, the size of the overall population usually remains about the same. Thus, an oak tree might produce many thousands of acorns each year, but few, if any, will survive to become full-grown trees.

Darwin—who conceived of his ideas in the 1830s but did not publish them until Wallace came to similar conclusions—presented the case for evolution in detail in his 1859 book *On the Origin of Species by Natural Selection*. Darwin proposed that there will be differences between offspring that survive and reproduce and those that do not. In particular, individuals that have heritable characteristics making them more likely to survive and reproduce in their particular environment will, on average, have a better chance of passing those characteristics on to their own offspring. In this way, as many generations pass, nature would select those individuals best suited to particular environments, a process Darwin called natural selection. Over very long times, Darwin argued, natural selection acting on varying individuals within a population of organisms could account for all of the great variety of organisms we see today, as well as for the species found as fossils.

If the central requirement of natural selection is variation within populations, what is the ultimate source of this variation? This problem plagued Darwin, and he never

From left, Charles Darwin (1809-1882), Alfred Russel Wallace (1823-1913), and Gregor Mendel (1822-1884) laid the foundations of modern evolutionary theory.

Glossary of Terms Used in Teaching About Evolution

Evolution: Change in the hereditary characteristics of groups of organisms over the course of generations. (Darwin referred to this process as "descent with modification.")

Species: In general, a group of organisms that can potentially breed with each other to produce fertile offspring and cannot breed with the members of other such groups.

Variation: Genetically determined differences in the characteristics of members of the same species.

Natural selection: Greater reproductive success among particular members of a species arising from genetically determined characteristics that confer an advantage in a particular environment.

found the answer, although he proposed some hypotheses. Darwin did not know that a contemporary, Gregor Mendel, had provided an important part of the solution. In his classic 1865 paper describing crossbreeding of varieties of peas, Mendel demonstrated that organisms acquire traits through discrete units of heredity which later came to be known as genes. The variation produced through these inherited traits is the raw material on which natural selection acts.

Mendel's paper was all but forgotten until 1890, when it was rediscovered and contributed to a growing wave of interest and research in genetics. But it was not immediately clear how to reconcile new findings about the mechanisms of inheritance with evolution through natural selection. Then, in the 1930s, a group of biologists demonstrated how the results of genetics research could both buttress and extend evolutionary theory. They showed that all variations, both slight and dramatic, arose through changes, or mutations, in genes. If a mutation enabled an organism to survive or reproduce more effectively, that mutation would tend to be preserved and spread in a population through natural selection. Evolution was thus seen to depend both on genetic mutations and on natural selection. Mutations provided abundant genetic variation, and natural selection sorted out the useful changes from the deleterious ones.

Selection by natural processes of favored variants explained many observations on the geography of species differences—why, for example, members of the same bird species might be larger and darker in the northern part of their range, and smaller and paler in the southern part. In this case, differences might be explained by the advantages of large size and dark coloration in forested, cold regions. And, if the species occupied the entire range continuously, genes favoring light color and small size would be able to flow into the northern population, and vice versa—prohibiting their separation into distinct species that are reproductively isolated from one another.

How new species are formed was a mystery that eluded biologists until information about genetics and the geographical distribution of animals and plants could be put together. As a result, it became clear that the most important source of new species is the process of geographical isolation—through which barriers to gene flow can be created. In the earlier example, the interposition of a major mountain barrier, or the origin of an intermediate desert, might create the needed isolation.

Other situations also encourage the formation of new species. Consider fish in a river that, over time, changes course so as to isolate a tributary. Or think of a set of oceanic islands, distant from the mainland and just far enough from one another that interchange among their populations is rare. These are ideal circumstances for creating reproductive barriers and allowing populations of the same species to diverge from one another under the influence of natural selection. After a time, the species become sufficiently different that even when reunited they remain reproductively isolated. They have become so different that they are unable to interbreed.

In the 1950s, the study of evolution entered a new phase. Biologists began to be able to determine the exact molecular structure of the proteins in living things—that is, the actual sequences of the amino acids that make up each protein. Almost immediately, it became clear that certain proteins that serve the same function in different species have very similar amino acid sequences. The protein evidence was completely consistent with the idea of a common evolutionary history for the planet's living things. Even more important, this knowledge provided important clues about the history of evolution that could not be obtained through the fossil record.

The discovery of the structure of DNA by Francis Crick and James Watson in 1953 extended the study of evolution to the most

Discovery of a Missing Link

As a zoologist I have discovered many phenomena that can be rationally explained only as products of evolution, but none so striking as the ancestor of the ants. Prior to 1967 the fossil record had yielded no specimens of wasps or other *Hymenopterous* insects that might be interpreted as the ancestors of the ants. This hypothetical form was a missing link of major importance in the study of evolution. We did have many fossils of ants dating back 50 million years. These were different species from those existing today, but their bodies still possessed the basic body form of modern ants. The missing link of ant evolution was often cited by creationists as evidence against evolution. Other ant specialists and I were convinced that the linking fossils would be found, and that most likely they would be associated with the late Mesozoic era, a time when many dinosaur and other vertebrate bones were fossilized but few insects. And that is exactly what happened. In 1967 I had the pleasure of studying two specimens collected in amber (fossilized resin) from New Jersey, and dating to the late Mesozoic about 90 million years ago. They were nearly exact intermediates between solitary wasps and the highly social modern ants, and so I gave them the scientific name *Sphecomyrma,* meaning "wasp ant." Since that time many more *Sphecomyrma* specimens of similar age have been found in the United States, Canada, and Siberia, but none belonging to the modern type. With each passing year, such fossils and other kinds of evidence tighten our conception of the evolutionary origin of this important group of insects.

—Edward O. Wilson

fundamental level. The sequence of the chemical bases in DNA both specifies the order of amino acids in proteins and determines which proteins are synthesized in which cells. In this way, DNA is the ultimate source of both change and continuity in evolution. The modification of DNA through occasional changes or rearrangements in the base sequences underlies the emergence of new traits, and thus of new species, in evolution. At the same time, all organisms use the same molecular codes to translate DNA base sequences into protein amino acid sequences. This uniformity in the genetic code is powerful evidence for the interrelatedness of living things, suggesting that all organisms presently alive share a common ancestor that can be traced back to the origins of life on earth.

One common misconception among students is that individual organisms change their characteristics in response to the environment. In other words, students often think that the environment acts on individual organisms to generate physical characteristics that can then be passed on genetically to offspring. But selection can work only on the genetic variation that already is present in any new generation, and genetic variation occurs randomly, not in response to the needs of a population or organism. In this sense, as Francois Jacob has written, evolution is a "tinkerer, not an engineer."[2] Evolution does not design new organisms; rather, new organisms emerge from the inherent genetic variation that occurs in organisms.

Genetic variation is random, but natural selection is not. Natural selection tests the combinations of genes represented in the members of a species and allows to proliferate those that confer the greatest ability to survive and reproduce. In this sense, evolution is not the simple product of random chance.

The booklet *Science and Creationism: A View from the National Academy of Sciences*[3] summarizes several compelling lines of evidence that demonstrate beyond any reasonable doubt that evolution occurred as a historical process and continues today. In brief:

• Fossils found in rocks of increasing age attest to the interrelated lineage of living things, from the single-celled organisms that lived billions of years ago to *Homo sapiens.* The most recent fossils closely resemble the organisms alive today, whereas increasingly older fossils are progressively different, providing compelling evidence of change through time.

• Even a casual look at different kinds of organisms reveals striking similarities among species, and anatomists have discovered that these similarities are more than skin deep. All vertebrates, for example, from fish to humans, have a common body plan characterized by a segmented body and a hollow main nerve cord along the back. The best available scientific explanation for these common structures is that all vertebrates are descended from a common ancestor species and that they have diverged through evolution.

• In the past, evolutionary relationships could be studied only by examining the consequences of genetic information, such as the anatomy, physiology, and embryology of living organisms. But the advent of molecular biology has made it possible to read the history of evolution that is written in every organism's DNA. This information has allowed organisms to be placed into a common evolutionary family tree in a much more detailed way than possible from previous evidence. For example, as described in Chapter 3, comparisons of the differences in DNA sequences among organisms provides evidence for many evolutionary events that cannot be found in the fossil record.

Evolution is the only plausible scientific explanation that accounts for the extensive array of observations summarized above. The concept of evolution through random genetic variation and natural selection makes sense of what would otherwise be a huge body of unconnected observations. It is no longer possible to sustain scientifically the view that the living things we see today did not evolve from earlier forms or that the human species was not produced by the same evolutionary mechanisms that apply to the rest of the living world.

The following two sections of this chapter examine two important themes in evolutionary theory. The first concerns the occurrence of evolution in "real time"—how changes come about and result in new kinds of species. The second is the ecological framework that underlies evolution, which is needed to understand the expansion of biological diversity.

Evolution as a Contemporary Process

Evolution by natural selection is not only a historical process—it still operates today. For example, the continual evolution

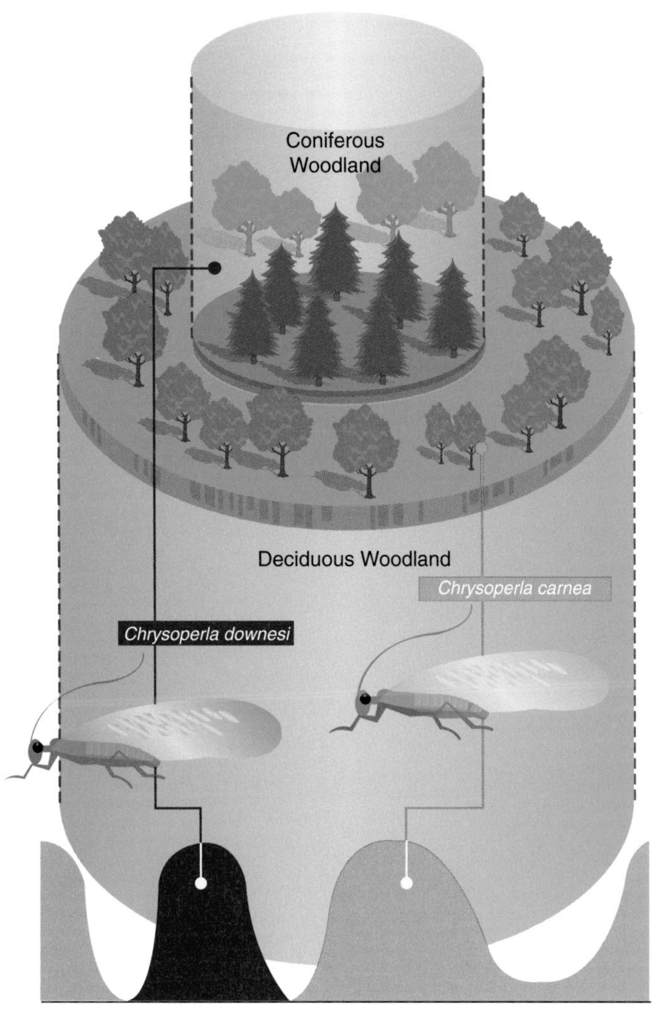

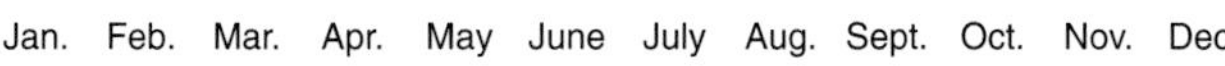

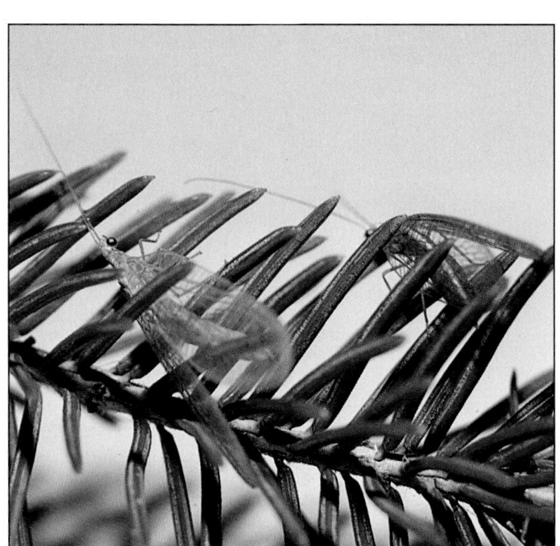

The North American lacewing species *Chrysoperla carnea* and *Chrysoperla downesi* separated from a common ancestor species recently in evolutionary time and are very similar. But they are different in color, reflecting their different habitats, and they breed at different times of the year.

of human pathogens has come to pose one of the most serious public health problems now facing human societies. Many strains of bacteria have become increasingly resistant to once-effective antibiotics as natural selection has amplified resistant strains that arose through naturally occurring genetic variation. The microorganisms that cause malaria, gonorrhea, tuberculosis, and many other diseases have demonstrated greatly increased resistance to the antibiotics and other drugs used to treat them in the past. The continued use and overuse of antibiotics has had the effect of selecting for resistant populations because the antibiotics give these strains an advantage over nonresistant strains.[4]

Similar episodes of rapid evolution are occurring in many different organisms. Rats have developed resistance to the poison warfarin. Many hundreds of insect species and other agricultural pests have evolved resistance to the pesticides used to combat them—and even to chemical defenses genetically engineered into plants. Species of plants have evolved tolerance to toxic metals and have reduced their interbreeding with nearby nontolerant plants—an initial step in the formation of separate species. New species of plants have arisen through the crossbreeding of native plants with plants introduced from elsewhere in the world.

The creation of a new species from a pre-existing species generally requires

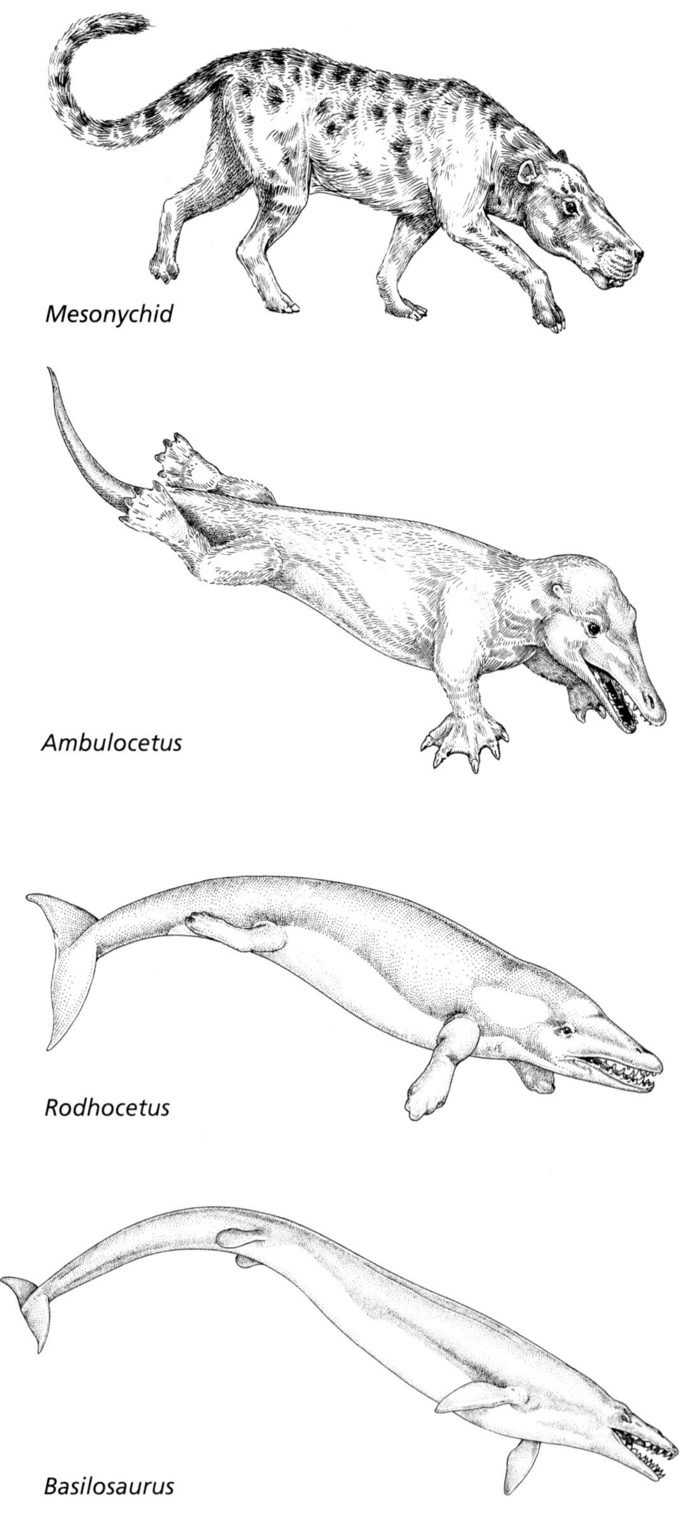

Modern whales evolved from a primitive group of hoofed mammals into species that were progressively more adapted to life in the water.

thousands of years, so over a lifetime a single human usually can witness only a tiny part of the speciation process. Yet even that glimpse of evolution at work powerfully confirms our ideas about the history and mechanisms of evolution. For example, many closely related species have been identified that split from a common ancestor very recently in evolutionary terms. An example is provided by the North American lacewings *Chrysoperla carnea* and *Chrysoperla downesi*. The former lives in deciduous woodlands and is pale green in summer and brown in winter. The latter lives among evergreen conifers and is dark green all year round. The two species are genetically and morphologically very similar. Yet they occupy different habitats and breed at different times of the year and so are reproductively isolated from each other.

The fossil record also sheds light on speciation. A particularly dramatic example comes from recently discovered fossil evidence documenting the evolution of whales and dolphins. The fossil record shows that these cetaceans evolved from a primitive group of hoofed mammals called *Mesonychids*. Some of these mammals crushed and ate turtles, as evidenced by the shape of their teeth. This mammal gave rise to a species with front forelimbs and powerful hind legs with large feet that were adapted for paddling. This animal, known as *Ambulocetus*, could have moved between sea and land. Its fossilized vertebrae also show that this animal could move its back in a strong up and down motion, which is the method modern cetaceans use to swim and dive. A later fossil in the series from Pakistan shows an animal with smaller functional hind limbs and even greater back flexibility. This species, *Rodhocetus*, probably did not venture onto land very often, if at all. Finally, *Basilosaurus* fossils from Egypt and the United States present a recognizable whale, with front flippers for steering and a completely flexible backbone. But this animal still has hind limbs (thought to have been nonfunctional),

Ongoing Evolution Among Darwin's Finches

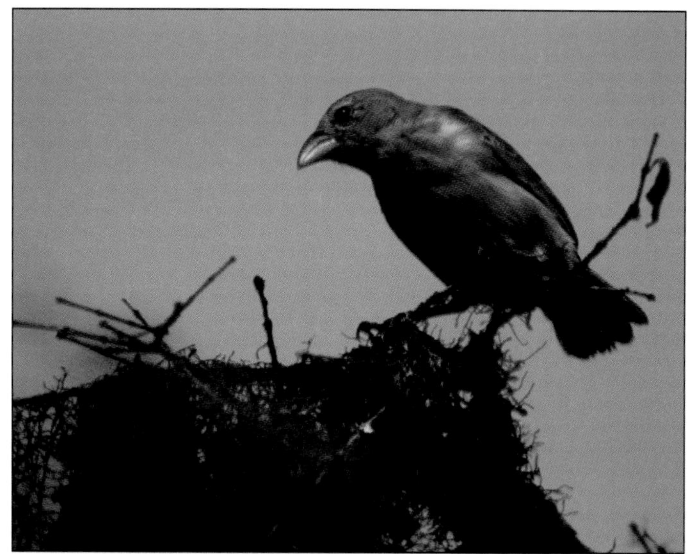

A particularly interesting example of contemporary evolution involves the 13 species of finches studied by Darwin on the Galapagos Islands, now known as Darwin's finches. A research group led by Peter and Rosemary Grant of Princeton University has shown that a single year of drought on the islands can drive evolutionary changes in the finches.[6] Drought diminishes supplies of easily cracked nuts but permits the survival of plants that produce larger, tougher nuts. Drought thus favors birds with strong, wide beaks that can break these tougher seeds, producing populations of birds with these traits. The Grants have estimated that if droughts occur about once every 10 years on the islands, a new species of finch might arise in only about 200 years.[7]

which have become further reduced in modern whales.[5]

Another focus of research has been the evolution of ancient apelike creatures through many intermediate forms into modern humans. *Homo sapiens*, one of 185 known living species in the primate order, is a member of the hominoids, a category that includes orangutans, gorillas, and chimpanzees. The succession of species that would give rise to humans seems to have separated from the succession that would lead to the apes about 5 to 8 million years ago. The first members of our genus, *Homo,* had evolved by about 1.5 million years ago. According to recent evidence—based on the sequencing of DNA found in a part of human cells known as mitochondria—it has been proposed that a small group of modern humans evolved in Africa about 150,000 years ago and spread throughout the world, replacing archaic populations of *Homo sapiens*.

Evolution and Ecology

Animals and plants do not live in isolation, nor do they evolve in isolation. Indeed, much of the pressure toward diversification comes not only from physical factors in the environment but from the presence of other species. Any animal is a potential host for parasites or prey for a carnivore. A plant has other plants as competitors for space and light, can be a host for parasites, and provides food for herbivores. The interactions within the complex communities, or ecosystems, in which organisms live can generate powerful evolutionary forces.

Evolution in natural communities arises from both constraints and opportunities. The constraints come from competitors, primarily among the same species. There are only so many nest holes for bluebirds and so much food for mice. Genetically different individuals that are able to move to a different resource—a new food supply, for example, or a hitherto uninhabited area—

Early hominids had smaller brains and larger faces than species belonging to the genus *Homo,* including our own species, *Homo sapiens*. White parts of the skulls are reconstructions, and the skulls are not all on the same scale.

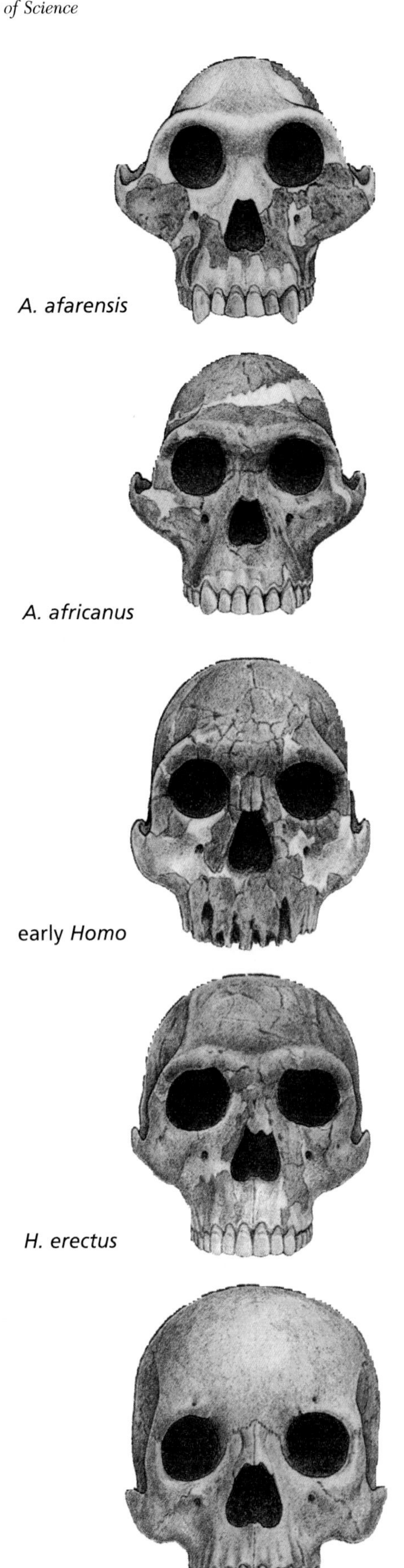

are able to exploit that resource free of competition. As a result, the trait that opened up the new opportunity will be favored by natural selection because the individuals possessing it are able to survive and reproduce better than other members of their species in the new environment.

An ecologist would say that the variant had occupied a new niche—a term that defines the "job description" of an organism. (For example, a bluebird would have the niche of insect- and fruit-eater, inhabitant of forest edges and meadows, tree-hole nester, and so on.) One often finds closely related species in the same place and occupying what look like identical niches. However, if the niches were truly identical, one of the species should have a competitive advantage over the other and eventually drive the less fit species to extinction or to a different niche. That leads to a tentative hypothesis: where we find such a situation, careful observation should reveal subtle niche specialization of the apparently competing species.

This hypothesis has been tested by many biologists. For example, in the 1960s Robert MacArthur carefully studied three North American warblers of the same genus that were regularly seen feeding on insects in coniferous trees in the same areas—indeed, often in the same trees. MacArthur's painstaking observations revealed that the three were actually specialists: one fed on insects on the major branches near the trunk; another occupied the mid-regions of branches and ate from different parts of the foliage; and the third fed on insects occupying the finest needles near the periphery of the tree. Although the three warblers occurred together, they were in fact not competitors for the same food resources.

Often, species that are evolving together in the same ecosystem do so through a highly interactive process. For example, natural selection will favor organisms with defenses against predation; in turn, predators experience selection for traits that overcome those defenses. Such coevolutionary competitions are common in nature. Many

A Chemical Distress Signal

J. H. Tumlinson and colleagues at the U.S. Department of Agriculture's Research Service Laboratories in Gainesville, Florida, have explored a fascinating case that illustrates the intricacy of many ecological relationships. Cotton plants, like many other crops, are attacked by caterpillars. One destructive cotton pest, the army worm, produces a complex series of reactions when it feeds on the plant—a reaction that involves the caterpillar itself, the tissues of the plant, and a third participant, a wasp that preys on the caterpillar. When the caterpillar chews on the cotton plant leaf, a reaction occurs that causes the plant to synthesize and release a class of volatile chemicals that escape into the air and travel rapidly downwind. The chemicals are detected by wasps, who follow the scent and are able to find the caterpillars and deposit eggs within them. The eggs hatch, and the wasp larvae destroy the caterpillar.[8]

This complex case of "chemical ecology" required a series of linked coevolutionary events: the response of the plant to a special signal from its predator, and the response of the wasp to a special signal from the host of its prey.

plants manufacture and store chemicals that deter herbivorous insects; but usually one or more insect species will have evolved biochemical mechanisms for inactivating the deterrent, providing them with a plant they can eat relatively free of competitors.

Another classic example of coevolution involves the introduction of rabbits and the myxomatosis virus into Australia. After rabbits were brought to Australia, they multiplied rapidly and threatened the wool industry because they grazed on the same plants as sheep. To control the rabbit population, a virulent pathogen of rabbits, the myxomatosis virus, also was introduced into Australia. Within a decade, rabbits had become more resistant to the virus, and the virus had evolved into a less virulent form, allowing both the host and pathogen to coexist.[9]

Conclusion

As the examples in this chapter demonstrate, evolutionary biology provides an extremely active and rich source of new insights into the world. By exploring the history of life on earth and shedding light on how evolution works, evolutionary biology is linking fundamental scientific research to knowledge needed to meet important societal needs, including the preservation of our environment. Few other ideas in science have had such a far-reaching impact on our thinking about ourselves and how we relate to the world.

Notes

1. Biological Sciences Curriculum Study. 1978. *Biology Teachers' Handbook*. 3rd ed. William V. Mayer, ed. New York: John Wiley and Sons.
2. Francois Jacob. June 10, 1977. Evolution and tinkering. *Science* 196:1161-1166.
3. National Academy of Sciences. (in press). *Science and Creationism: A View from the National Academy of Sciences*. Washington, DC: National Academy Press. (See www.nap.edu)
4. P. Ewald. 1994. *The Evolution of Infectious Disease*. New York: Oxford University Press.
5. "Evolution, Science, and Society: A White Paper on Behalf of the Field of Evolutionary Biology," Draft, June 4, 1997.
6. Jonathan Weiner. 1994. *The Beak of the Finch*. New York: Alfred A. Knopf.
7. Peter R. Grant. 1991. Natural selection and Darwin's finches. *Scientific American*, October, pp. 82-87.
8. James H. Tumlinson, W. Joe Lewis, and Louise E. M. Vet. 1993. How parasitic wasps find their hosts. *Scientific American*, March, pp. 100-106.
9. F. Fenner and F.N. Ratcliffe. 1965. *Myxomatosis*. Cambridge: Cambridge University Press.

Dialogue

Teaching About the Nature of Science

In the following vignette, Barbara, Doug, and Karen use a model to continue their discussion of the nature of science and its implications for the teaching of evolution.

....

"Thanks for meeting with me this afternoon," Barbara says. "To begin this demonstration I first need to ask you what you think science is."

"Oh, I had that in college," says Karen. "The scientific method is to identify a question, gather information about it, develop a hypothesis that answers the question, and then do an experiment that either proves or disproves the hypothesis."

"But that was one of my points about evolution," Doug says. "No one was there when evolution happened and we can't do any experiments about what happened in the past. So by your definition, Karen, evolution isn't science."

"Science is a lot more than just supporting or rejecting hypotheses," Barbara replies. "It also involves observation, creativity, and judgment. Here's an activity I use to teach the nature of science."

Barbara takes a cardboard mailing tube about one foot long that has the ends of four ropes extending from it.

As Barbara tugs on the various ropes one at a time, she has Doug and Karen make observations of what happens. After three or four pulls, she asks Karen and Doug to predict what will happen when she pulls on one of the ropes. Both are able to predict that if Barbara pulls rope A, rope B will move. Barbara then asks if there are additional manipulations they would like to see, and she follows their requests.

Barbara then asks Doug and Karen to sketch a model of what is inside the tube that could explain their observations.

When Karen and Doug show their sketches to each other, they realize that they have come up with different models. Barbara asks them if they want to make any changes to their sketches based on the comparison, and both of them make modifications, although their final models are still different.

Barbara then gives them their own cardboard tubes and some string and asks them to build the model they proposed. When their models are built, Barbara holds up her tube and asks Doug and Karen to follow her actions with their own models, to see if the two models behave in the same way as Barbara's tube. But when Barbara pulls string A in her tube, Karen's model does not work the same way. Karen asks if she can make some changes in her model, and once she does her new model seems to work the same way as Barbara's. Doug's model consistently behaves the same way as Barbara's tube.

"Now wait a minute," Karen says. "What do ropes and tubes have to do with science and evolution?"

"You might not know it, but what we just did is much of what science is about. You observed what happened when

I pulled these ropes. Then, based on your initial observations, you made a prediction about what would happen if we manipulated the system in a specific way. How accurate was your prediction?"

"We were right," Doug responds.

"And why were you able to predict what would happen before I pulled the rope?"

"I used what I observed in the first few pulls to help me predict what would happen later."

"Basically what each of you did was to speculate about how my tube was working on the basis of some limited observations. Scientists do that type of thing all the time. They make observations and try to explain what's going on, or sometimes they recognize that more than one explanation fits their data. Then they try out their proposed explanations by making predictions that they test. At first I had you draw a picture of how you thought my tube worked and had you each explain your picture. You got to hear each other's view on how the system worked. Doug, did you change your ideas at all based on what you heard from Karen?"

"Well, yes. I first thought that ropes A and C were the two ends of the same rope and B and D were two ends of another rope. Karen had A and B as ends of the same rope and C and D as ends of another rope, and her explanation seemed to fit better than mine."

"Right. Communication about observations and interpretations is very important among scientists because different scientists may interpret data in different ways. Hearing someone else's views can help a scientist revise his or her interpretation. In essence that was what you were doing when you shared your diagrams. Karen, when your model didn't work, what did you do?"

"All I did was adjust the length of one rope, and then it worked fine."

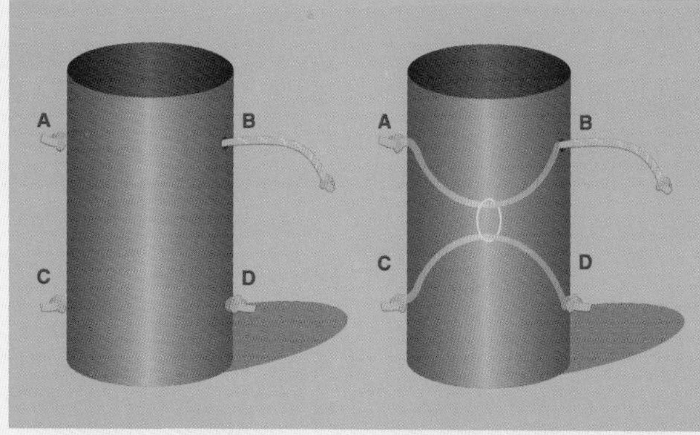

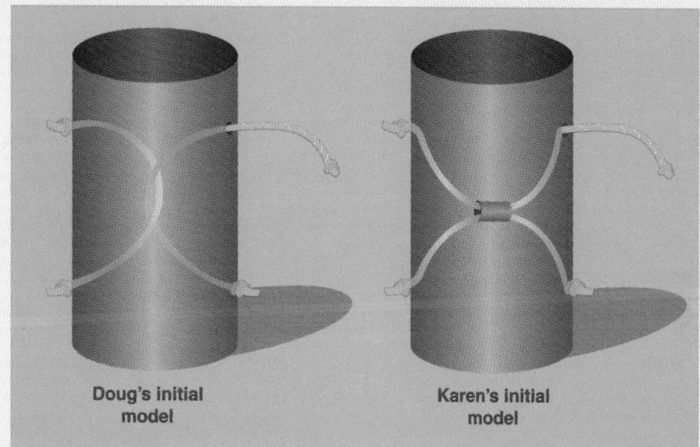

"So as a result of your formal testing of the predictions from your model, you revised your explanation of the system. Your understanding improved. In scientific terms, you revised your model to make it more consistent with your further observations. In science, the validity of any explanation is determined by its coherence with observations in the natural world and by its ability to predict further observations."

"But we still have different models," Karen observes. "How do we know which one is right?"

Doug says: "You told us that, didn't you, Barbara. There can be two possible explanations for the same observation."

"So it's possible for scientists to disagree sometimes," says Karen. "But does that mean that we don't understand evolution because scientists disagree about how evolution takes place?"

"Not at all," Barbara answers, "you both created different models of my tube, but both of your models are fairly accurate. And don't forget there were constraints on the possible models you could create that would be consistent with the data. Just any explanation would not be acceptable. In evolution, there are some things we know could not have happened, just as we are confident that some things have happened."

"And if different scientists can have different explanations, like Karen and I did, then I guess science also has to involve judgment to some extent," Doug says.

"But I thought scientists were supposed to be totally objective," says Karen.

"Good science always attempts to be objective, but it also relies on the individual insights of scientists. And the questions they choose to ask as well as the methods they choose to use, not to mention the interpretations they may have, can be colored by their individual interests and backgrounds. But scientific explanations are reviewed by other scientists and must be consistent with the natural world and future experiments, so there are checks on subjectivity. What we read in science books is a combination of observations and inferred explanations of those observations that can change with new research."

"Still, I'm wondering," says Karen, "how can we find out which model is right?"

"Let's just open up Barbara's tube," says Doug.

"We could do that," Barbara says. "But let's assume in this analogy that opening the tube is not possible. Sometimes scientists figure out how to open up the natural world and look inside, but sometimes they can't. And not opening up the tube is a good metaphor for how science often works. Science involves coming up with explanations that are based on evidence. With time, additional evidence might require changing the explanations, so that at any time what we have is the best explanation possible for how things work. In the future, with additional data, we may change our original explanation—just like you did, Karen.

"Remember when we were talking this morning about evolution being fact or theory? That conversation is very relevant to what we have been doing with the tubes. As scientists started to notice patterns in nature, they began to speculate about some explanations for these patterns. These explanations are analogous to your initial ideas about how my tube worked. In the terms of science, these initial ideas are called hypotheses. You noticed some patterns in how the ropes were related to each other, and you used these patterns to develop a model to explain the patterns. The model you created is analogous to the beginning of a scientific theory. Except in science, theories are only formalized after many years of testing the predictions that come from the model.

"Because of our human limitations in collecting complete data, theories necessarily contain some judgments about what is important. Judgments aren't a weakness of scientific theory. They are a basic part of how science works."

"I always thought of science as a bunch of absolute facts," says Doug. "I never thought about how knowledge is developed by scientists."

"Creativity and insight are what help make science such a powerful way of understanding the natural world.

"There's another important thing that I try to teach my students with this activity," Barbara continues. "It's important for them to be able to distinguish questions that can be answered by science from those that cannot be answered by science. Here's a list of questions that I use to get them talking. I ask them if a question can be answered by science, cannot be answered by science, or has some parts that belong to science and others that do not. Then I ask the group to select a couple of questions and discuss how they would go about answering them."

Barbara hands Doug and Karen the following list of questions:

Do ghosts haunt old houses at night?
How old is the earth?
Should I follow the advice of my daily horoscope?
Do species change over long periods of time?
Should I exercise regularly?

"Of course, you can make up other questions if something is happening in the news or if it's related to an earlier lesson. And sometimes I include moral or religious questions to make it clear that they lie outside science."

"I can see that these would get students thinking," says Karen. "I guess understanding the nature of science really is relevant to real life."

"That's what this exercise is about."

TROPICVS CANCRI
CAPRICORNVS
AQVARIVS
SAGITTARIVS
REGVNT ORBEM
LIBYA
Garama
SINVS

3

Evolution and the Nature of Science

Science is a particular way of knowing about the world. In science, explanations are restricted to those that can be inferred from confirmable data—the results obtained through observations and experiments that can be substantiated by other scientists. Anything that can be observed or measured is amenable to scientific investigation. Explanations that cannot be based on empirical evidence are not a part of science.

The history of life on earth is a fascinating subject that can be studied through observations made today, and these observations have led to compelling accounts of how organisms have changed over time. The best available evidence suggests that life on earth began more than three and a half billion years ago. For more than two billion years after that, life was housed in the bodies of many kinds of tiny, single-celled organisms, some of which produced the oxygen that now makes up more than a fifth of the earth's atmosphere. Less than a billion years ago, much more complex organisms appeared. By about half a billion years ago, evolution had resulted in a wide variety of multicellular animals and plants living in the sea that are the clear ancestors of many of the major types of organisms that continue to live to this day. Somewhat more than 400 million years ago, some marine plants and animals began one of the greatest of all innovations in evolution—they invaded dry land. For our own phylum, the Chordata, this move away from the nurturing sea led to the appearance of amphibians, reptiles, birds, and mammals—the latter including, of course, our own species, *Homo sapiens*.

This chapter looks at how science works in the context of our overall understanding of how biological evolution occurred. It begins, however, by discussing another scientific development that challenged long-held understandings and beliefs: the discovery of heliocentricism.

Heliocentricism and the Nature of Science

Surely one of the first major natural phenomena to be understood was the cause of night and day. Some of the earliest surviving human records left on clay tablets relate to the movements of the sun and other celestial bodies. The obvious cause of day and night is the rising and setting of the sun. This is an observation that can be made today by anyone and, seemingly, requires no further explanation.

Archaeological evidence and early records make it clear that our ancestors realized that not only does the sun appear to rise and set, but so do the moon and stars. The movements of the moon and stars, however, are not precisely synchronized with

Clockwise from top left, Nicolaus Copernicus (1473-1543), Johannes Kepler (1571-1630), Galileo Galilei (1564-1642), and Isaac Newton (1642-1727) led the way to a new understanding of the relationship between the earth and the sun and initiated an age of scientific progress that continues today.

Illustration from the 18th century depicts the Ptolemaic system in the upper left corner and the Copernican system in other corners and center.

those of the sun. The moon is slower by about one hour per day. The stars remain almost the same on successive nights, but slowly it becomes obvious that they, too, are slowed in their movements compared to the sun. Thus, the stars of summer are different from those visible in the winter. In fact, it takes a full year for the stars to return to their previous position, an interval of time that defines our year.

The ancient observers realized that not all stars move in unison. Although most move in majestic unity, a few others are "wanderers"—appearing now with one group of stars and a week later somewhere else. The majority were called "fixed stars," the wanderers were called "planets."

During the late Middle Ages, and especially in the Renaissance, beautiful brass models known as orreries were made to show the relative positions and movements of the sun, planets, and moon as they circled the earth. As the center of the universe, the earth was a sphere in the center of the orrery. The other celestial bodies were positioned on rings of metal, each moving by clockwork at its own rate. The fixed stars required a simple solution—they could be considered stuck in an outermost shell, also moved by clockwork.

The problem with orreries—and with the theories of the cosmos then prevailing—was that they had to become successively more complex as more became known. Careful observations of the movements of the stars and planets greatly complicated the hypotheses used to account for those movements. This growing complexity stimulated some of the leading astronomers of the 16th and 17th centuries, including Copernicus, Kepler, and Galileo, to make even more precise observations of the movements of the heavenly bodies. Astronomers used these measurements to demonstrate that the age-old human explanations of the heavens were incomplete. In the process, they replaced a complex and confusing explanation with a simple one: the sun, rather than the earth, is at the center of a "solar system," and the earth revolves around it. That simple step—a bold departure from past thinking due mainly to the insights of Copernicus (1473 to 1543)—dramatically changed the picture of the then known universe.

This concept of heliocentricism initially ran counter to the positions of religious authorities. The view of Christianity over most of its history, based on a literal interpretation of the Bible, was that the earth is the center of the universe around which the celestial bodies revolve. Copernicus dedicated his book describing the theory of heliocentricism, *De revolutionibus orbium coelestium*, to Pope Paul III and promptly died. That saved him the troubles that were to beset Galileo (1564 to 1642), whose astronomical observations confirmed the views of Copernicus. Galileo was told to abandon his beliefs, and he later was tried by the Inquisition and sentenced to the equivalent of house arrest. The Church held that his views were dangerous to faith.

Continued study and ever more careful measurements of the movements of the planets and sun continued to support the heliocentric hypothesis. Then, in the latter half of the 17th century, Isaac Newton (1642 to 1727) showed that the force of gravity—as measured on the earth—could account for the movements of the planets given the laws of motion that Newton derived. As a result of the steady accumulation of evidence, the theological interpretation of celestial movements gave way to the naturalistic explanation, and it is now accepted that night and day are the consequences of the rotation of the earth on its axis. Today, we can see for ourselves the rotation of the earth from satellites orbiting the planet.

Like biological evolution, the theory of heliocentricism brought order and new understanding to an otherwise chaotic and confusing aspect of nature. It also had great practical applications, in that the exploration of the world by European seafarers used the more accurate understanding of celestial mechanics to assist in navigation.

Looking at the night sky remains a powerful experience. But that experience is now informed not only by the beauty and majesty of the heavens, but by a deeper understanding of nature and by an appreciation of the power of the human intellect.

This triumph of the human mind says a great deal about the nature of science. First, science is not the same as common sense. Common sense indicates that the sun does rise and set. Nevertheless, there can be other explanations of that phenomenon, and one of them, the rotation of the earth on its axis, is responsible for day and night. A concept based on observation proved to need extensive modification as new observations accumulated.

Second, the statements of science should never be accepted as "final truth." Instead, over time they generally form a sequence of increasingly more accurate statements. Nevertheless, in the case of heliocentricism as in evolution, the data are so convincing that the accuracy of the theory is no longer questioned in science.

Third, scientific progress depends on individuals, but the contributions of one individual could be made by others. If Copernicus had kept his ideas to himself, the discovery of heliocentricism would have been postponed, but it would not have been blocked, since other astronomers eventually would have come to the same conclusion.

Biologists have used construction cranes to study the many newly discovered species that live in the canopies of tropical forests, as in this research project in Panama.

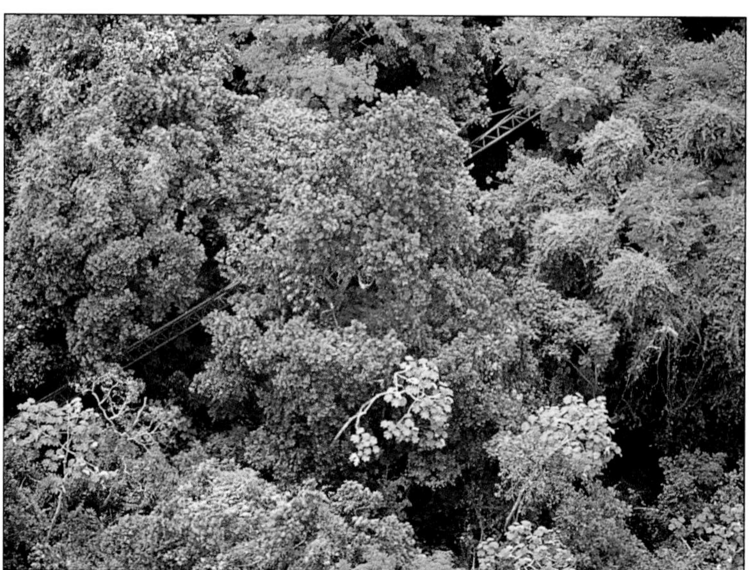

Similarly, had Darwin and Wallace not published their hypotheses, the concept of biological evolution would nevertheless have emerged as the accepted explanation for the history of life on earth. The same cannot be said in other areas of human endeavor; for example, had Shakespeare never published, we would most assuredly never have had his plays. The publications of scientists, unlike those of playwrights, are a means to an end—they are not the end itself.

Science Requires Careful Description

What are the scientific methods that have led to our current understanding of the history of life over vast eons of time? They begin with careful descriptions of the material being studied.

The material for the study of biological evolution is life itself. One basic aspect of life is that individuals can be grouped as similar kinds, or species. Another important observation is that many species seem to be closely related to each other. The scientific classification of species and their arrangement into groups began with the publication in 1758 of *Systema Naturae*, or system of nature, by the Swedish naturalist Carolus Linnaeus (1707 to 1778). For example, Linnaeus knew seven dog-like species, and he gave each a double name. Subsequently many more species were discovered and some of the names were changed—and continue to be changed as more information is obtained. The domestic dog is *Canis familiaris*; the coyote of North America is *Canis latrans*; the Australian dingo is *Canis dingo*; and the wolf of the northern hemisphere is *Canis lupus*. Thus *Canis* is the name of the genus of dog-like animals, and the distinctive second name is the species name.

Generations of scientists have discovered new species, described them, and arranged them into the system first suggested by Linnaeus. Whereas Linnaeus recognized about 9,000 species, systematists now have recognized about 1.5 million. The task of categorizing and describing species is still far from complete. Most species of smaller invertebrates, and many bacteria and other microscopic organisms, remain to be discovered. The plant kingdom is also incompletely known. Though the flowering plants of many areas, such as Europe and North America, are fairly well described, many other regions have not been nearly as well explored by botanists.

Recent investigations in the exceptionally diverse rainforests of South America have caused biologists to raise their estimates of the number of undescribed species. For example, a very high proportion of insects collected from the forest canopy are "new" species to science. It is now believed that the number of different species of plants and animals in the world may be ten million, or even more.

The scientific methods used in classifying organisms have been greatly improved over time. The process begins with the intensive field work in which the animals, plants, and microorganisms are collected and carefully examined. Most will be known to a specialist, but there might be some unusual examples. However, none is likely to be a complete stranger, since the specialist will probably recognize that any puzzling specimen is similar to some familiar species. Next the specialist must check all that has been published on the group of organisms that contains the similar species. If, after an exhaustive search, there is no record of a described species that corresponds to the one being examined, the specimen is probably a new species. The specialist will then prepare a careful description of the new species and publish it in a scientific journal. There is a permanent reward for being the describer of a new species: thereafter monographs that deal with the classification of the group to which the new species belongs will add the describer's name at the end of the scientific name. Thus, for example, "*Homo sapiens* Linnaeus" is our own proper identification, because Linnaeus was the first to give us our scientific name.

Despite their similarities with birds, bats are mammals that evolved from flightless ancestors.

This example makes it clear that not all scientific data are derived as the result of experimentation. The conventional classification of species into seemingly natural groups involved the careful observation of a variety of different species, followed by the use of selected characteristics in an attempt to define groups of species thought to be related. But the groupings are not always obvious. For example, it might have seemed reasonable to classify bats with birds, since the most conspicuous characteristic of each is the ability to fly. But bats are mammals. Like all mammals, their bodies are covered with hair and their young are born alive (instead of hatching from eggs) and are nourished by milk from the mother's mammary glands.

Although most of the species we know today were described after the time of Linnaeus, we continue to use his basic system of hierarchical classification. For example, similar genera are united in families, similar families in orders, similar orders in classes, and similar classes in phyla. The dog-like species listed above (the genus *Canis*), plus a number of similar but more distant dog-like animals, are grouped as the family Canidae. This family plus the families of cats, bears, seals, and weasels form the order Carnivora. The carnivores and all other animals with hair are combined as the class Mammalia. Mammals are combined with the birds, reptiles, amphibians, fishes, plus some small marine animals in the phylum Chordata. Today, many systematists group organisms according to a system known as cladistics. By determining which traits of a species evolved earlier and which evolved later, this system seeks to classify organisms according to their evolutionary history.

Science as Explanation

In the quest for understanding, science involves a great deal of careful observation that eventually produces an elaborate written description of the natural world. This description is communicated to scientists in scientific journals or at scientific meetings, so that others can build on pre-existing work. In this way, the accuracy and sophistication of the description tends to increase with time, as subsequent generations of scientists correct and extend the observations of their predecessors. Because the total sum of scientific knowledge increases relentlessly, scientific progress is something that all scientists take for granted.

But science is not just description. Even as observations are being made, the human mind attempts to sort, or organize, the observations in a way that reveals some underlying order in the objects or phenomena being observed. This sorting process, which involves a great deal of trial and error, seems to be driven by a fundamental human urge to make sense of our world.

The sorting process also suggests new observations that might otherwise not be made. For example, the suggestion that bats should be grouped with mammals led to an intensified examination of the similarities between bats and rodents—first at the anatomical level, and later with respect to the genes and protein molecules that form their cells. In this case, new evidence was obtained that confirmed the suggested relationship. In other cases, the further observations inspired by a tentative grouping have caused the rejection of a new idea.

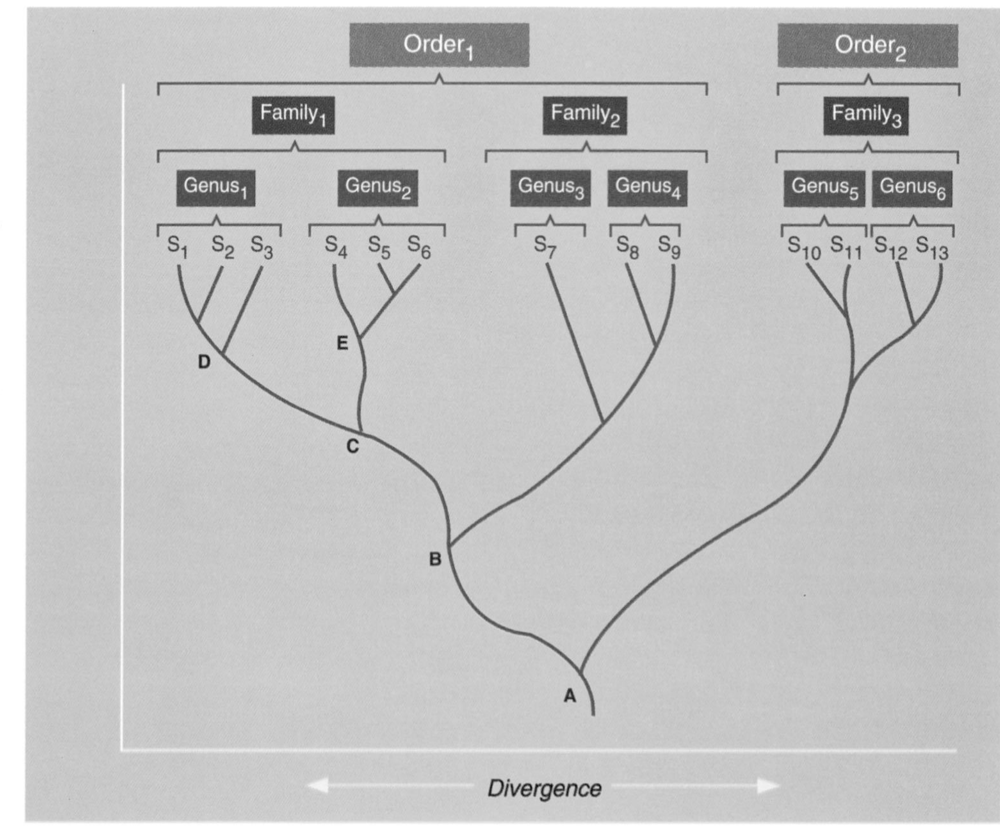

Evolutionary relationships often are depicted in diagrams that resemble the branches of trees. Closely related species (denoted S1, S2, etc.) are grouped into genera, genera into families, and so on. The result is a hierarchical diagram showing how different species evolved from common ancestor species (represented in this diagram by the letters A through E).

The realization that species can be arranged in a hierarchy of groups of seemingly similar forms raised an obvious question: What accounts for the relatedness of different groups of organisms? The mechanism that was proposed by Darwin directly addressed this question. It suggested that all animals classified as belonging to the same group had a common ancestor species. That is, dogs, wolves, coyotes and all members of the genus *Canis* are descended from a common ancestor species that lived in the remote past. In a similar manner all species in a family, an order, a class, or a phylum share a common inheritance.

How could one possibly test such a hypothesis? In the decades before Darwin proposed his hypothesis, geologists realized that the sedimentary rocks of the earth's crust contain a running diary of earth's history. This record of past events comes about because the earth's crust is in a constant state of change. This observation might not be obvious in the lifetime of an individual, but it is dramatic over thousands of years. Relatively flat surfaces are uplifted to form mountains, and then the mountains slowly erode to form flatlands. Storms produce powerful waves that erode cliffs at the seashore. These phenomena have the common feature of moving solid materials, and the subsequent settling out of these materials makes possible the formation of a special form of rock that contains a record of the earth's past.

Consider the case of a river with a source in the mountains. As the water moves downstream, it erodes the slopes of the mountains. Tiny grains produced by the erosion, called silt, are relatively easy to move. When the river reaches the flatlands, a lake, or the ocean, the solid material being carried by the water is deposited—often reaching great thicknesses over long periods of time. Then the pressure of the sediments on top can cause the sediments beneath to harden into "sedimentary rocks."

The river may carry things other than silt, sand, and rocks. Hard structures of organisms such as the bones and teeth of animals may be carried along as well. These, too, will be deposited with the silt, sand, and rocks. Under certain circumstances, these remains of organisms undergo a chemical change in which the original material is replaced by molecules that form stone. In this way, the organic remains of living things are fossilized (changed into stone), creating the evidence of ancient life studied by scientists.

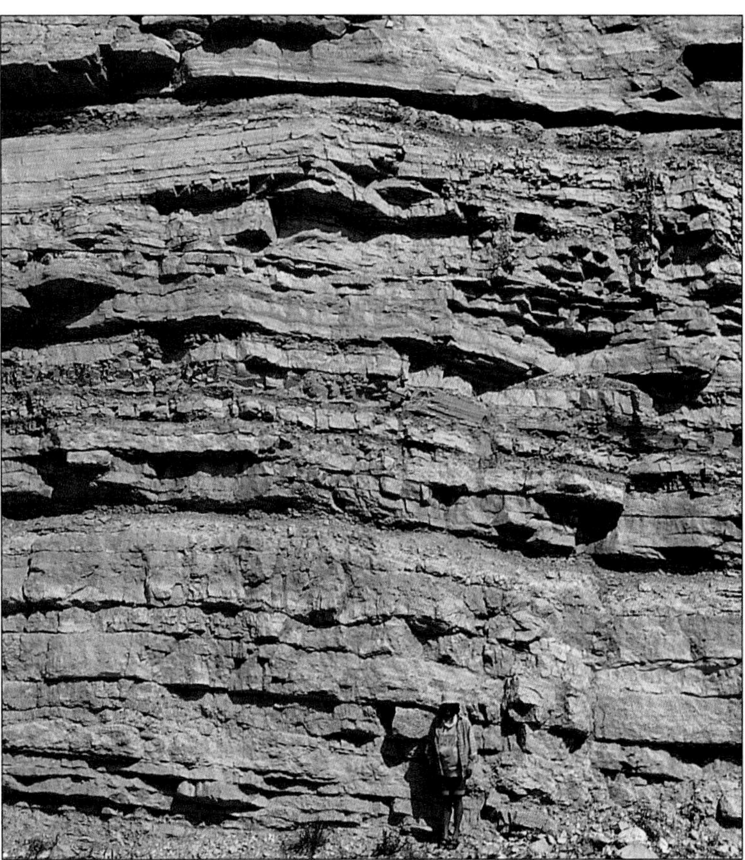

Sedimentary rocks are formed when solid materials carried by wind or water accumulate in layers and then are compressed by overlying deposits. Sedimentary rocks sometimes contain fossils formed from the parts of organisms deposited along with other solid materials.

Because of the order in which the sediments are deposited, the most recent layer of rocks normally will be on top and the oldest layer will be on the bottom (though sometimes sediments are flipped upside down by the geologic folding of rock layers). Also, the fossils in each layer usually will be of those organisms that lived at the time the layer was formed. Thus, the fossils in the lower layers will represent species that lived earlier than those found in the upper layers.

The relative position of fossils tells only which are older and which younger. One

In the fossil record, most species are characterized by a specific appearance, a duration over time, and extinction. The evolutionary origins of species are inferred from the morphological relations among fossils.

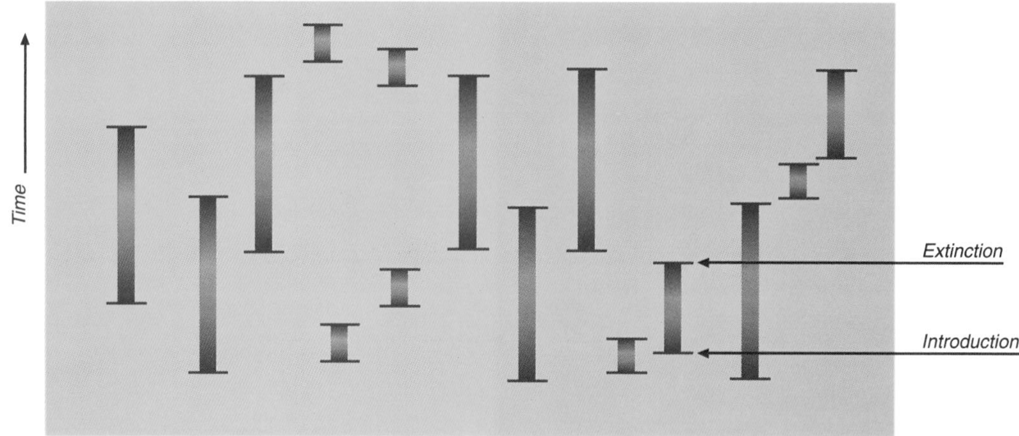

Types of fossils in fossil record

can estimate the difference in the ages of the two fossils by noting the thickness of the rock that separates them. If the difference is only one foot, one might guess the interval of time is less than if two fossils are separated by 50 feet of rock layers. Today, however, far more accurate methods of dating fossils are available, as described on the next page. Because these methods are based on the known rates of radioactive decay, they provide valuable measures of absolute time.

The scientific study of fossils is called paleontology, and the methods used for their identification and classification are similar to those used for living species. But in some respects the task of the paleontologist is far more difficult. Many species lack hard parts such as bones and shells, and such organisms almost always decay without becoming fossilized. This is the case for many groups of soft-bodied invertebrates—such as worms of many kinds, jellyfish, and protozoans. Even for such species as mammals, birds, reptiles, and amphibians, death is usually followed by the skeleton being dismembered and the bones scattered. For this reason, whereas isolated bones are often fossilized, it is exceptionally rare for an intact skeleton to be found.

Tiny fossils first reveal the existence of bacteria 3.5 to 3.8 billion years ago, and animals composed of more than a single cell are known from about 670 million years ago. But the organisms that lived between these two dates lacked hard parts and, hence, were rarely preserved as fossils. Then, about 570 million years ago, a dramatic change took place. At the beginning of the Cambrian period, animals evolved that had calcified shells and other types of body coverings that had a far better chance of becoming fossilized. These fossils demonstrate that Cambrian seas were populated with a variety of invertebrates. The earliest vertebrate fossils date from about 500 million years ago. Thereafter early amphibians and reptiles appeared. Birds and mammals appear in the fossil record only about 200 million years ago, while dinosaurs first appear about 225 million years ago and disappear suddenly about 160 million years later.

In the 1830s, when Darwin began his studies, the essential features of the fossil record were known (although absolute dates had not yet been determined). Many thousands of living species had been described, and it was clearly recognized that they could be organized into various groups—suggesting that they are somehow relatives. In addition, analysis of the fossil record revealed that the organisms on the ancient earth had undergone major changes over time—with whole groups of animals appearing, persisting for long periods of time, and then disappearing.

Darwin was an unusually keen observer. But he was not content to catalogue facts and observations. Instead, the natural world to him was a gigantic, very challenging puzzle that demanded an explanation for its otherwise bewildering complexity. Why are different organisms so similar? Why has there been a succession of different kinds of species throughout geologic time?

Certain observations seemed particularly important. For example:

1) In South America, the only continent where living armadillos were found, Darwin discovered fossil evidence for the prior existence of ancient species that had many of the unique features of living armadillos, yet were clearly different. Such fossils were found nowhere else in the world. Why were both living and ancient armadillo-like species confined to the same geographical region?

2) On the Galapagos Islands, 600 miles off the coast of Ecuador, Darwin observed many distinct living species of birds and reptiles that closely resembled each other—yet were different on each tiny island. Why, for example, should the beak size of the

Dating the Earth

One of the greatest scientific triumphs of the last two centuries has been the discovery of the vast expanse of geologic time. Early methods of calculating the age of the earth relied on measures of the rate of sedimentation or the cooling of the earth from an initially molten state. The relative ages of rocks also were calculated early in the 1800s by noting what kinds of fossils the rocks contained. But the absolute age of the earth and the timing of many events in geologic history required the discovery late in the 19th century of a previously unknown phenomenon: radioactivity.

Some elements, such as uranium, undergo radioactive decay to produce other elements. By measuring the quantities of radioactive elements and the elements into which they decay in rocks, geologists can determine how much time has elapsed since the rock cooled from an initially molten state. For example, the oldest known rocks are found in Greenland and date from about 3.8 billion years ago. Scientists believe the earth's age to be about 4.6 billion years because meteorites and rocks of the moon—both of which formed about the same time as the earth—date from this time. Radiometric dating also shows that the period of earth's history during which large fossils can be readily found in rocks began only about 570 million years ago.

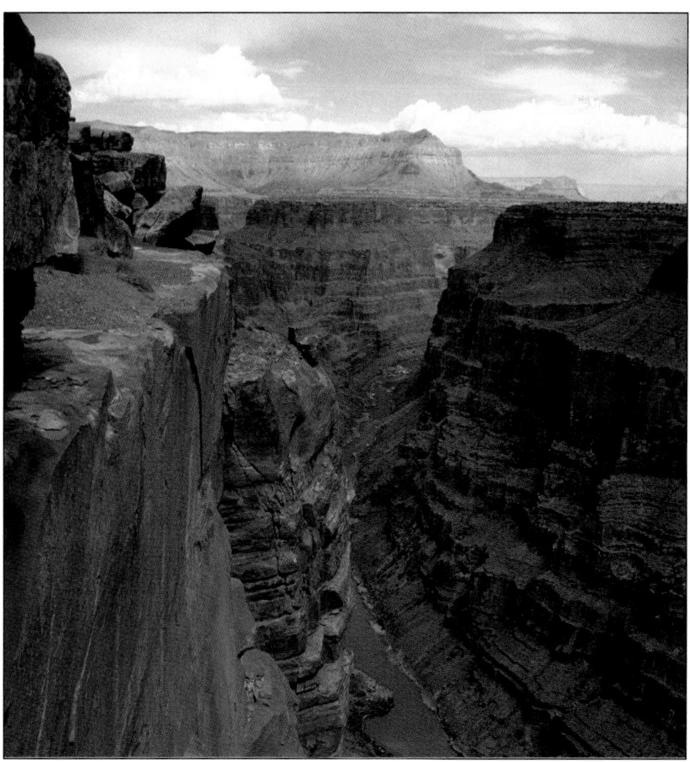

Radiometric dating draws on information and insights from many areas of science. For example, it requires that the rate of radioactive decay is constant over time and is not influenced by such factors as temperature or pressure—conclusions supported by extensive research in physics. It also assumes that the rocks being analyzed have not been altered over time by the migration of atoms into or out of the rocks, which requires detailed information from both the geologic and chemical sciences.

In South American, Darwin found fossil species that were clearly related to modern armadillos, yet neither the fossils nor the living animals were found anywhere else in the world. In *The Origin of Species*, he explained that "the inhabitants of each quarter of the world will obviously tend to leave in that quarter closely allied though modified descendants."

Before the start of the Cambrian period about 550 million years ago, multicellular organisms lacked hard parts like shells and bones and rarely left fossils. However, a few pre-Cambrian organisms left traces of their existence. Some ancient rocks contain stromatolites—the remnants of bacteria that grew in columns like stacked pancakes (right). Above, a fossil just predating the Cambrian shows the outlines of a marine invertebrate that might have resembled a jellyfish.

A timeline of evolution demonstrates the tremendous expanse of geologic time compared to the period since humans evolved. Each higher scale details part of the scale beneath it. While the estimated times of various evolutionary events continue to change as new fossils are discovered and dating methods are refined, the overall sequence demonstrates both the scope and grandeur of evolutionary change.

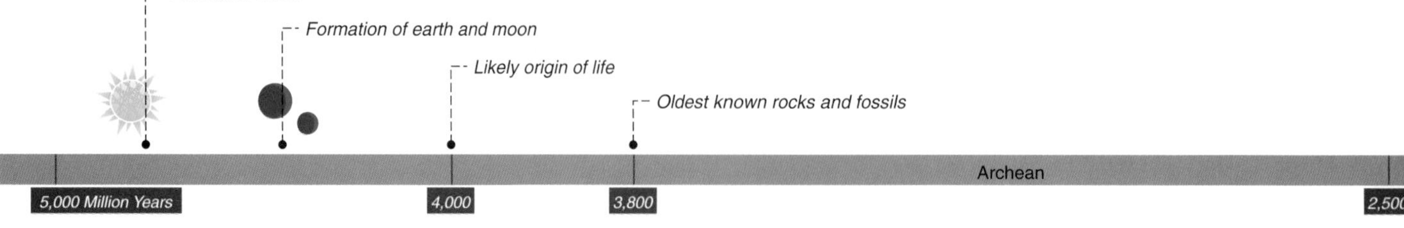

mockingbirds on one island be different from that of a closely related mockingbird on an island only 30 miles away? And why were the various types of animals on these islands related, but distinct from, the animals in Ecuador, whereas those on the otherwise very similar islands off the coast of Africa were related to the animals in Africa instead?

Darwin could not see how these observations could be explained by the prevailing view of his time: that each species had been independently created, with the species that were best suited to each location on the earth being created at each particular site. It looked instead as though species could evolve from one into another over time, with each being confined to the particular geographical region where its ancestors happened to be—particularly if isolated by major barriers to migration, such as vast expanses of ocean.

But how could one species turn into another over the course of time? In constructing his hypothesis of how this occurred, Darwin was struck by several other observations that he and others before him had made.

1) People who bred domesticated animals and plants for commercial or recreational use had found and exploited a great deal of variation among the progeny of their crosses. Pigeon breeders, for example, had observed wide differences in colors, beaks, necks, feet, and tails of the offspring from a single mating pair. They routinely enhanced their stocks for desired traits—for example, selectively breeding those animals that shared a particular type of beak. Through such artificial selection, pigeon fanciers had been able to create many different-looking pigeons, known as breeds. A similar type of artificial selection

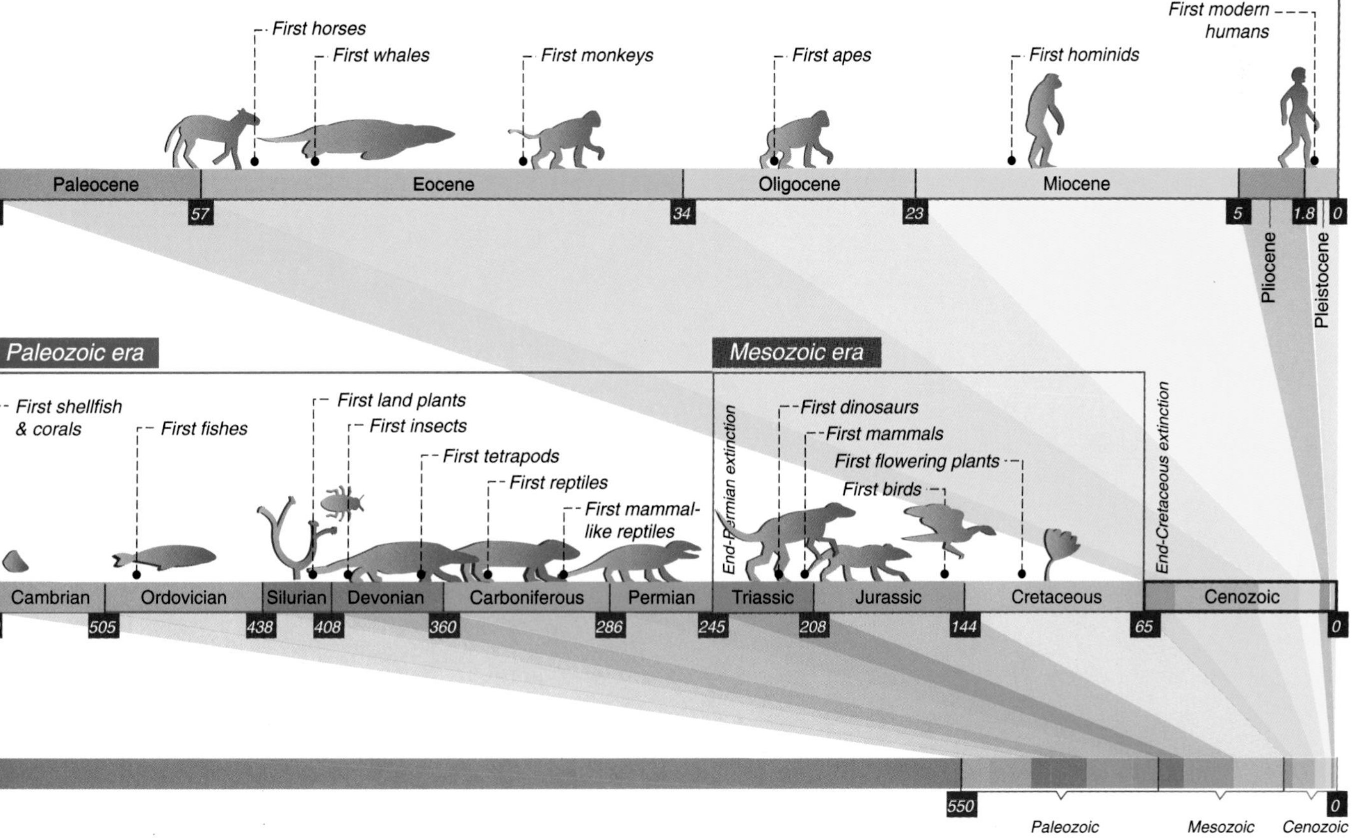

for mating pairs of dogs had likewise created the whole variety of shapes and sizes of these common pets—ranging from a Great Dane to a dachshund.

2) Animals living in the wild can face a tremendous struggle for survival. For some birds, for example, fewer than one in 100 animals born in one year will survive over a harsh winter into a second year. Those with characteristics best suited for a particular environment—for example, those individual birds who are best able to find scarce food in the winter while avoiding becoming food for a larger animal—tend to have better chances of surviving. Darwin called this process natural selection to distinguish it from the artificial selection used by dog and pigeon breeders to determine which animals to mate to produce offspring.

At least 20 years elapsed between the time that Darwin conceived of descent with modification and 1859, the year that he revealed his ideas to the world in *On the Origin of Species*. Throughout these 20 years, Darwin did what scientists today do: he tested his ideas of how things work with new observations and experiments. In part, he did this by thinking up every possible objection he could to his own hypothesis. For each such argument, Darwin tried to find an observation made by others, make an observation, or do an experiment of his own that might imply that his ideas were in fact not valid. When he could successfully counter such objections, he strengthened his theory. For example, Darwin's ideas readily explained why distant oceanic islands were generally devoid of terrestrial mammals, except for flying bats. But how could the land snails, so common on such islands, have traversed the hundreds of miles of open ocean that separate the islands from the mainland where the snails first evolved? By floating snails on salt-water for prolonged periods, Darwin convinced himself that, on rare occasions in the past, snails might in fact have "floated in chunks of drifted timber across moderately wide arms of the sea."

This example shows how a hypothesis can drive a scientist to do experiments that would otherwise not be done. Prior to Darwin, the existence of land snails and bats, but not typical terrestrial mammals, on the oceanic islands was simply noted and catalogued as a fact. It is unlikely that anyone would have thought to test the snails for their ability to survive for prolonged periods in salt water. Even if they had, such an experiment would have had little meaning or impact.

The ability to analyze individual biological molecules has added great detail to biologists' understanding of the tree of life. For example, molecular analyses indicate that all living things fall into three domains—the Bacteria, Archaea, and Eucarya—related by descent from a common ancestor.

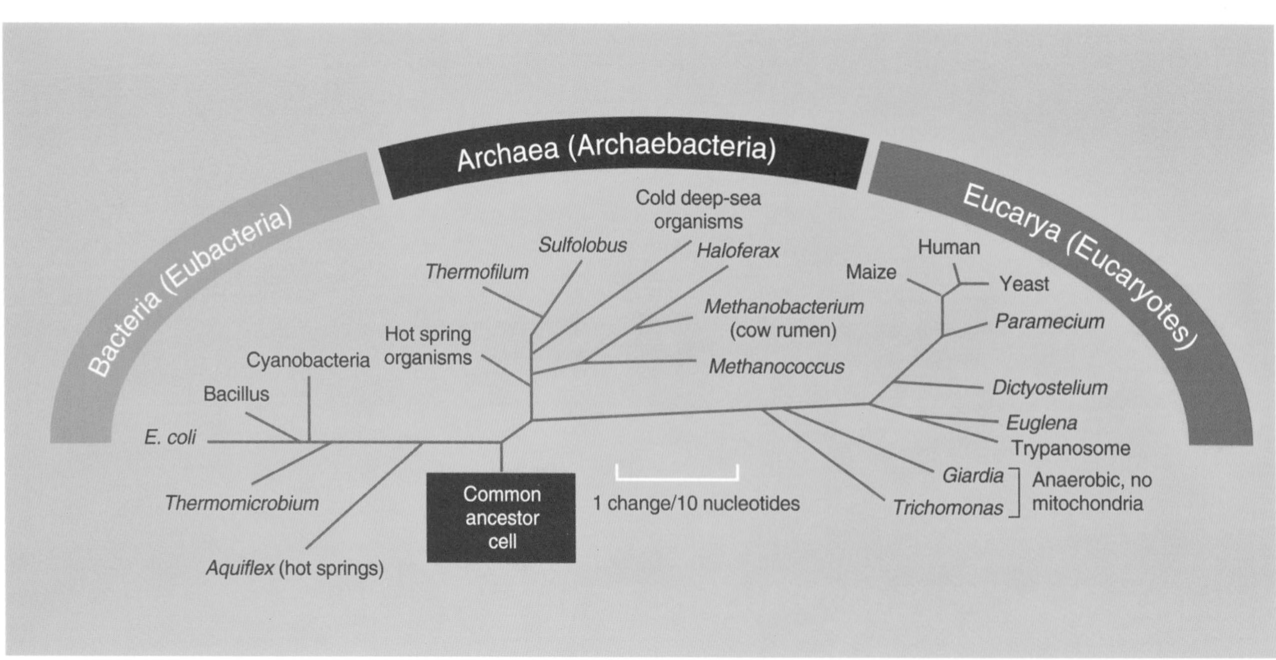

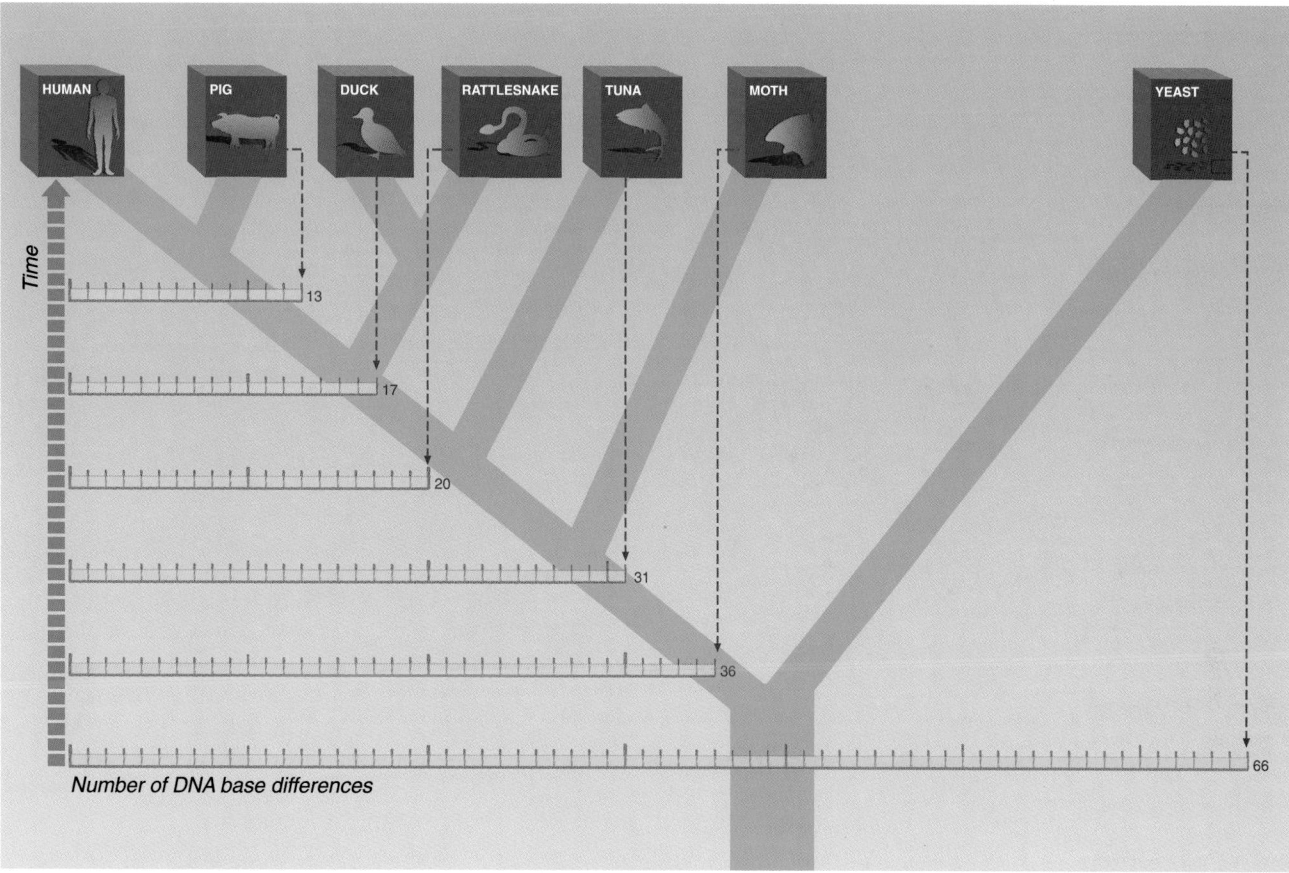

Organisms ranging from yeast to humans use an enzyme known as cytochrome C to produce high-energy molecules as part of their metabolism. The gene that codes for cytochrome C gradually has changed over the course of evolution. The greater the differences in the DNA bases that code for the enzyme, the longer the time since two organisms shared a common ancestor. This DNA evidence for evolution has confirmed evolutionary relationships derived from other observations.

By publishing his ideas, Darwin subjected his hypothesis to the tests of others. This process of public scrutiny is an essential part of science. It works to eliminate individual bias and subjectivity, because others must also be able to determine whether a proposed explanation is consistent with the available evidence. It also leads to further observations or to experiments designed to test hypotheses, which has the effect of advancing science.

Many of the hypotheses advanced by scientists turn out to be incorrect when tested by further observations or experiments. But skillful scientists like Darwin tend to have good ideas that end up increasing the amount of knowledge in the world. For this reason, the ideas of scientists have been—over the long run—central to much of human progress.

Science as Cumulative Knowledge

At the time of Darwin, there were many unsolved puzzles, including missing links in the fossil record between major groups of animals. Guided by the central idea of evolution, thousands of scientists have spent their lives searching for evidence that either supports or conflicts with the idea. For example, since Darwin's time, paleontologists have discovered many ancient organisms that connect major groups—such as *Archaeopteryx* between ancient reptiles and birds, and *Ichthyostega* between ancient fish and amphibians. By now, so much evidence has been found that supports the fundamental idea of biological evolution that its occurrence is no longer questioned in science.

Even more striking has been the information obtained during the 20th century from studies on the molecular basis of life. The theory of evolution implies that each organism should contain detailed molecular evidence of its relative place in the hierarchy of living things. This evidence can be found in the DNA sequences of living organisms. Before a cell can divide to produce two daughter cells, it must make a new copy of its DNA. In copying its DNA nucleotides, however, cells inevitably make a small number of mistakes. For this reason, a few nucleotides are changed through random error each time that a cell divides. (For example, an A in the DNA sequence of a gene in a chromosome may be replaced with a G in the new copy made as the cell divides.) Therefore, the larger number of cell divisions that have elapsed between the time that two organisms diverged from their common ancestor, the more differences there will be in their DNA sequences due to chance errors.

This molecular divergence allows researchers to track evolutionary events by sequencing the DNA of different organisms. For example, the lineage that led to humans and to chimpanzees diverged about 5 million years ago—whereas one needs to look back in time about 80 million years to find the last common ancestor shared by mice and

Continental Drift and Plate Tectonics:

A Scientific Revolution of the Past 50 Years

The theory of plate tectonics demonstrates that revolutions in science are not just a thing of the past, thus suggesting that more revolutions can be expected in the future.

World maps have long indicated a curious "jigsaw puzzle fit" of the continents. This is especially apparent between the facing coastlines of South America and Africa. Alfred Wegener (1880 to 1930), a German meteorologist who was dissatisfied with explanations that relied on expanding and contracting crust to account for mountain building and the formation of the ocean floor, pursued other lines of reasoning. Wegener suggested that all of earth's continents used to be assembled in a single ancient super-continent he called Pangea. He hypothesized that Pangea began to break up approximately 200 million years ago, with South America and Africa slowly drifting apart to their present positions, leaving the southern Atlantic Ocean between them. This was an astonishing hypothesis: could huge continents really move?

Wegener cited both geological and biological evidence in support of his explanation. Similar plant and animal fossils are found in rock layers more than 200 million years old in those regions where he claimed that different continents were once aligned. Wegener attributed this to the migration of plants and animals freely throughout these broad regions. If 200 million years ago Africa and South America had been separated by the Atlantic Ocean as they are today, their climates, environments, and life forms should have been very different from each other—but they were not.

Despite Wegener's use of evidence and logic to develop his explanations, other scientists found it difficult to imagine how solid, brittle continents could plow through the equally solid and brittle rock material of the ocean floor. Wegener did not have an explanation for how the continents moved. Since there was no plausible mechanism for continental drift, the idea did not take hold. The hypothesis of continental drift was equivalent to the hypothesis of evolution in the decades before Darwin, when evolution lacked the idea of variation followed by natural selection as an explanatory mechanism.

The argument essentially lay dormant until improved technologies allowed scientists to gather previously unobtainable data. From the mid 1950s through the early 1970s, new evidence for a mechanism to explain continental drift became available that the scientific community could accept. Sonar mapping of the ocean floor revealed the presence of a winding, continuous ridge system around the globe. These ridges were places where molten material was welling up from the earth's interior and pushing apart the plates that form the earth's surface.

humans. As a result, there is a much smaller difference between human and chimpanzee DNA than between human (or chimpanzee) and mouse DNA. In fact, scientists today routinely use the differences they can measure between the DNA sequences of organisms as "molecular clocks" to decipher the relationships between living things.

The same comparisons among organisms can be made using the proteins encoded by DNA. For example, every living cell uses a protein called cytochrome c in its energy metabolism. The cytochrome c proteins from humans and chimpanzees are identical. But there is only an 86 percent overlap in the molecules between humans and rattlesnakes, and only a 58 percent overlap between us and brewer's yeast. This is explained by the evolutionary proposition that we shared a common ancestor with chimps relatively recently, whereas the common ancestor that we, as vertebrates, shared with rattlesnakes is much more ancient. Still farther in the past, we and yeast shared a common ancestor—and the molecular data reflect this pattern.

In the past few decades, new methods have been developed that are allowing us to

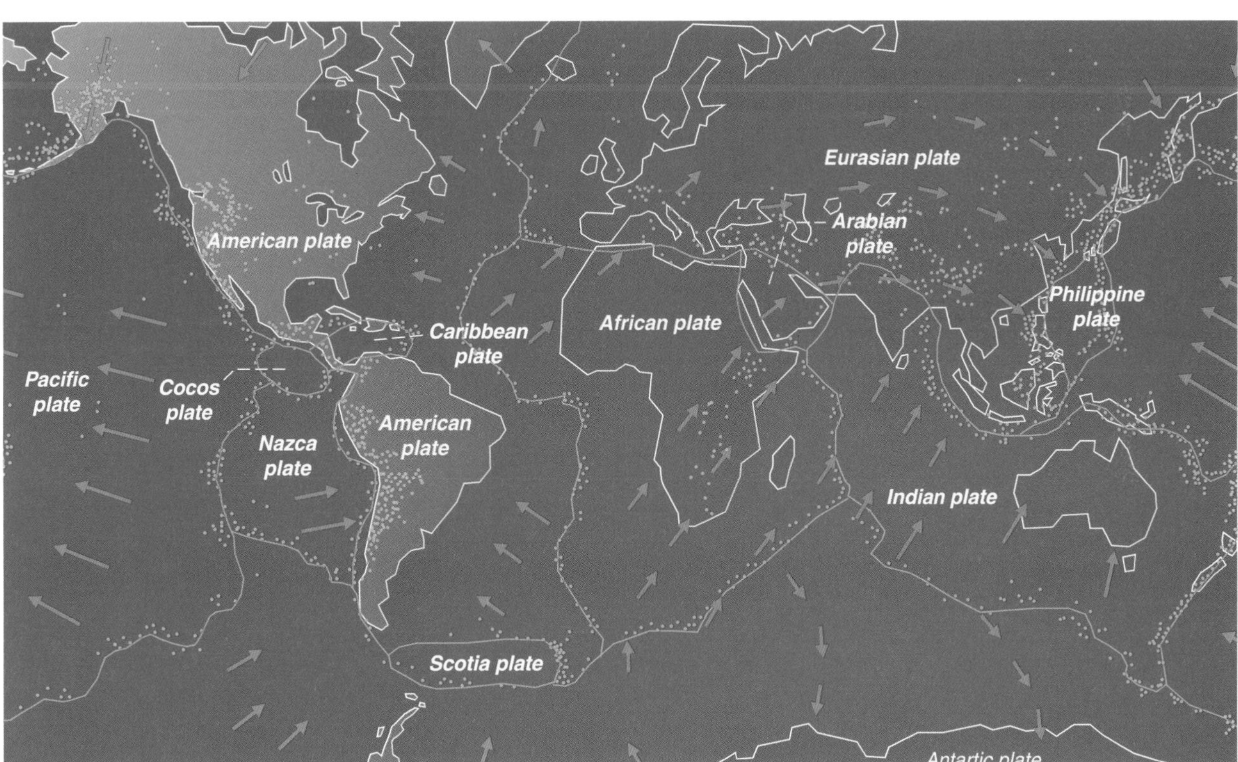

In a relatively short time, these new observations, measurements, and interpretations provoked a complete shift in the thinking of the scientific community. Geologists now accept the idea that the surface of the earth is broken up into about a dozen large pieces, as well as a number of smaller ones, called tectonic plates.

On a time scale of millions of years, these plates shift about on the planet's surface, changing the relative positions of the continents. The plate tectonic model provides explanations that are widely accepted for the evolution of crustal features such as folded mountain chains, zones of active volcanoes and earthquakes, and deep ocean floor trenches. Direct measurements using the satellite-based global positioning system (GPS) to measure absolute longitude and latitude verify that the plates collide, move apart, and slide past one another in different areas along their adjacent boundaries at speeds comparable to the growth rate of a human fingernail.

obtain the exact sequence of all of the DNA nucleotides in chromosomes. The Human Genome Project, for example, will produce when completed the entire sequence of the 3 billion nucleotides that make up our genetic inheritance. The complete sequence of the yeast genome (12 million nucleotides) is already known, as are the genomes for numerous species of bacteria (from 0.5 to 5 million nucleotides each, depending on the species). Similar sequencing efforts will soon yield the complete sequences for hundreds of bacteria and other organisms with small genomes.

These molecular studies are powerful evidence for evolution. The exact order of the genes on our chromosomes can be used to predict the order on monkey or even mouse chromosomes, since long stretches of the chromosomes of mammalian species are so similar. Even the parts of our DNA that do not code for proteins and at this point have no known function are similar to the comparable parts of DNA in related organisms.

The confirmation of Darwin's ideas about "descent with modification" by this recent molecular evidence has been one of the most exciting developments in biology in this century. In fact, as the chromosomes of more and more organisms are sequenced over the next few decades, these data will be used to reconstruct much of the missing history of life on earth—thereby compensating for many of the gaps that still remain in the fossil record.

Conclusion

One goal of science is to understand nature. "Understanding" in science means relating one natural phenomenon to another and recognizing the causes and effects of phenomena. Thus, scientists develop explanations for the changing of the seasons, the movements of heavenly bodies, the structure of matter, the shaping of mountains and valleys, the changes in the positions of continents over time, and the diversity of living things.

The statements of science must invoke only natural things and processes. The statements of science are those that emerge from the application of human intelligence to data obtained from observation and experiment. These fundamental characteristics of science have demonstrated remarkable power in allowing us to describe the natural world accurately and to identify the underlying causes of natural phenomena. This understanding has great practical value, in part because it allows us to better predict future events that rely on natural processes.

Progress in science consists of the development of better explanations for the causes of natural phenomena. Scientists can never be sure that a given explanation is complete and final. Yet many scientific explanations have been so thoroughly tested and confirmed that they are held with great confidence.

The theory of evolution is one of these explanations. An enormous amount of scientific investigation has converted what was initially a hypothesis into a theory that is no longer questioned in science. At the same time, evolution remains an extremely active field of research, with an abundance of new discoveries that are continually increasing our understanding of exactly how the evolution of living organisms actually occurred.

THE CONCERNS OF SCIENCE

An Excerpt from the Book

This Is Biology: The Science of the Living World (1997)

By Ernst Mayr

It has been said that the scientist searches for truth, but many people who are not scientists claim the same. The world and all that is in it are the sphere of interest not only of scientists but also of theologians, philosophers, poets, and politicians. How can one make a demarcation between their concerns and those of the scientist?

How Science Differs from Theology

The demarcation between science and theology is perhaps easiest, because scientists do not invoke the supernatural to explain how the natural world works, and they do not rely on divine revelation to understand it. When early humans tried to give explanations for natural phenomena, particularly for disasters, invariably they invoked supernatural beings and forces, and even today divine revelation is as legitimate a source of truth for many pious Christians as is science. Virtually all scientists known to me personally have religion in the best sense of this word, but scientists do not invoke supernatural causation or divine revelation.

Another feature of science that distinguishes it from theology is its openness. Religions are characterized by their relative inviolability; in revealed religions, a difference in the interpretation of even a single word in the revealed founding document may lead to the origin of a new religion. This contrasts dramatically with the situation in any active field of science, where one finds different versions of almost any theory. New conjectures are made continuously, earlier ones are refuted, and at all times considerable intellectual diversity exists. Indeed, it is by a Darwinian process of variation and selection in the formation and testing of hypotheses that science advances.

Despite the openness of science to new facts and hypotheses, it must be said that virtually all scientists—somewhat like theologians—bring a set of what we might call "first principles" with them to the study of the natural world. One of these axiomatic assumptions is that there is a real world independent of human perceptions. This might be called the principle of objectivity (as opposed to subjectivity) or common-sense realism. This principle does not mean that individual scientists are always "objective" or even that objectivity among human beings is possible in any absolute sense. What it does mean is that an objective world exists outside of the influence of subjective human perception. Most scientists—though not all—believe in this axiom.

Second, scientists assume that this world is not chaotic but is structured in some way, and that most, if not all, aspects of this structure will yield to the tools of scientific investigation. A primary tool used in all scientific activity is testing. Every new fact and every new explanation must be tested again and again, preferably by different investigators using different methods. Every confirmation strengthens the probability of the "truth" of a fact or explanation, and every falsification or refutation strengthens the probability that an opposing theory is correct. One of the most characteristic features of science is this openness to challenge. The willingness to abandon a currently accepted belief when a new, better one is proposed is an important demarcation between science and religious dogma.

The method used to test for "truth" in science will vary depending on whether one is testing a fact or an explanation. The existence of a continent of Atlantis between Europe and America became doubtful when no such continent was discovered during the first few Atlantic crossings in the period of discoveries during the late fifteenth and early sixteenth centuries. After complete oceanographic surveys of the Atlantic Ocean were made and, even more convincingly, after photographs from satellites were taken in this century, the new evidence conclusively proved that no such continent exists. Often, in science, the absolute truth of a fact can be established. The absolute truth of an explanation or theory is much harder, and usually takes much longer, to gain acceptance. The "theory" of evolution through natural selection was not fully accepted as valid by scientists for over 100 years; and even today, in some religious sects, there are people who do not believe it.

Third, most scientists assume that there is historical and causal continuity among all phenomena in the material universe, and they include within the domain of legitimate scientific study everything known to exist or to happen in this universe. But they do not go beyond the material world. Theologians may also be interested in the physical world, but in addition they usually believe in a metaphysical or supernatural realm inhabited by souls, spirits, angels, or gods, and this heaven or nirvana is often believed to be the future resting place of all believers after death. Such supernatural constructions are beyond the scope of science.

Dialogue

TEACHING EVOLUTION THROUGH INQUIRY

The following dialogue demonstrates a way of teaching about evolution using inquiry-based learning. High school students are often interested in fossils and in what fossils indicate about organisms and their habitats. In the investigation described here, the students conduct an inquiry to answer an apparently simple question: What influence has evolution had on two slightly different species of fossils? The investigation begins with a straightforward task—describing the characteristics of two species of brachiopods.

"Students, I want you to look at some fossils," says Karen. She gives the students a set of calipers and two plastic sheets that each contain about 100 replicas of carefully selected fossil brachiopods.[1] "These two sheets contain fossils from two different species of a marine animal called a brachiopod. Let's begin with some observations of what they look like."

"They look like butterflies," replies one student.

"They are kind of triangular with a big middle section and ribs," says another student.

"Can you tell if there are any differences between the fossils in the two trays?"

The students quickly conclude that the fossils have different sizes but that they cannot really tell any other difference.

"In that case, how could you tell if the fossil populations are different?" Karen asks.

"We can count the ribs."

"We can measure them."

"Those are both good answers. Here's what I want you to do. Break into groups of four and decide among yourselves which of those two characteristics of the fossils you want to measure. Then graph your measurements for each of the two different populations."

For the rest of the class period, the students investigate the fossils. They soon realize that the number of ribs is related to the size of the fossils, so the groups focus on measuring the lengths and widths of the fossils. They enter the data on the two different populations into a computer data

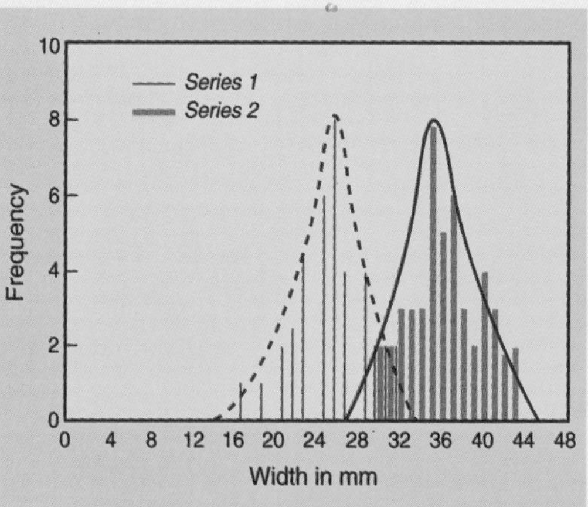

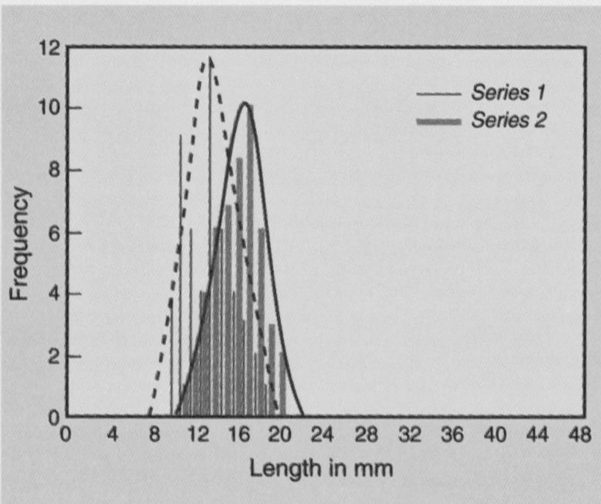

Graphs showing characteristics of brachiopod populations.

base. Two of the graphs that they generate are shown on the facing page.

"Now that we have these graphs of the fossils' lengths and widths," Karen says at the beginning of the next class period, "we can begin to talk about what these measurements mean. We see from one set of graphs that the fossils in the second group tend to be both wider and longer than those in the other group. What could that mean?"

"Maybe one group is older," volunteers one of the students.

"Maybe they're different kinds of fossils," says another.

"Let's think about that," says Karen. "How could their lengths and widths have made a difference to these organisms?"

"It could have something to do with the way they moved around."

"Or how they ate."

"That's good," says Karen. "Now, if you had dug up these fossils, you would have some additional information to work with, so let me give you some of that background. As I mentioned last week, these fossils are from marine animals known as brachiopods. When they die their shells are often buried in sediments and fossilized. What I know about the fossils you have is that they were taken from sediments that are about 400 million years old. But the two sets of fossils were separated in time by about 10 million years.

"Taking that information, I'd like you to do some research on brachiopods and develop some hypotheses about whether or not evolution has influenced their size. Here are some of the questions you can consider as you're writing up your arguments."

Karen hands out a sheet of paper containing the following questions:

- What differences in structure and function might be represented in the length and width of the brachiopods? Could efficiency in burrowing or protection against predators have influenced their shapes?

- Why might natural selection influence the lengths and widths of brachiopods?

- What could account for changes in their dimensions?

The following week, Karen holds small conferences at which the students' papers are presented and discussed. She focuses students on their ability to ask skeptical questions, evaluate the use of evidence, assess the understanding of geological and biological concepts, and review aspects of scientific inquiries. During the discussions, students are directed to address the following questions: What evidence would you look for that might indicate these brachiopods were the same or different species? How could changes in their shapes have affected their ability to reproduce successfully? What would be the likely effects of other changes in the environment on the species?

Note

1. The materials needed to carry out this investigation are available from Carolina Biological Supply Company, 2700 York Rd., Burlington, NC 27215. Phone: 1-800-334-5551. www.carolina.com

4

Evolution and the *National Science Education Standards*

Over the last six years, several major documents have been released that describe what students from kindergarten through twelfth grade should know and be able to do as a result of their instruction in the sciences. These include the *National Science Education Standards* released by the National Research Council in 1996,[1] the *Benchmarks for Science Literacy* released by the American Association for the Advancement of Science in 1993,[2] and *The Content Core: A Guide for Curriculum Designers* released by the Scope, Sequence, Coordination project of the National Science Teachers Association in 1992.[3]

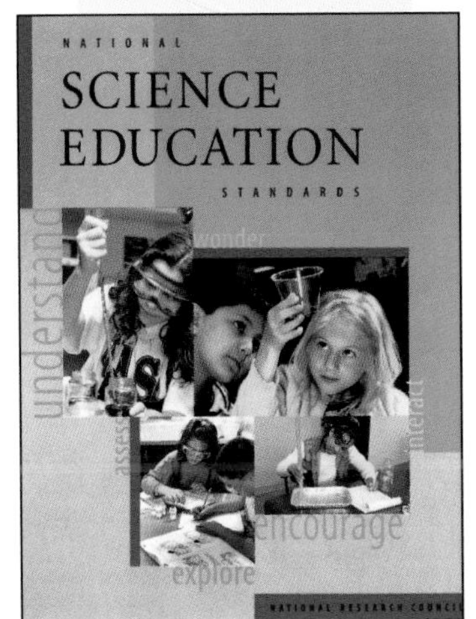

These documents agree that all students should leave biology class with an understanding of the basic concepts of biological evolution and of the limits, possibilities, and dynamics of science as a way of knowing. *Benchmarks for Science Literacy*, for example, states that "the educational goal should be for all children to understand the concept of evolution by natural selection, the evidence and arguments that support it, and its importance in history." For biology educators, these documents offer significant support for the inclusion of evolution in school science programs.

Structure and Overview of the *National Science Education Standards*

This chapter focuses on the treatment of evolution in the *National Science Education Standards*. The *Standards* are divided into six broad sections. The first set of standards, the *science teaching standards*, describes what teachers of science at all grade levels should know and be able to do. The *professional development standards* describe the experiences necessary for teachers to gain the knowledge, understanding, and ability to implement the *Standards*. The *assessment standards* provide criteria against which to judge whether assessments are contributing fully to the goals outlined in the *Standards*. The *science content standards* outline what students should know, understand, and be able to do in the natural sciences. The *science education program standards* discuss the planning and actions needed to translate the *Standards* into programs that reflect local contexts and policies. And the *science education system standards* consist of criteria for judging the performance of the overall science education system.

The *Standards* rest on the premise that science is an active process. Learning science is something that students do, not something that is done to them. "Hands-on" activities, although essential, are not enough. Students must have "minds-on" experiences as well.

The *Standards* make inquiry a central part of science learning. When engaging in inquiry, students describe objects and events, ask questions, construct explanations, test those explanations against current scientific knowledge, and communicate their ideas to others. They identify their assumptions, use critical and logical thinking, and consider alternative explanations. In this way, students actively develop their understanding of science by combining scientific knowledge with reasoning and thinking skills.

The importance of inquiry does not imply that all teachers should pursue a single approach to teaching science. Just as inquiry has many different facets, so too do teachers need to use many different strategies to develop the understandings and abilities described in the *Standards*.

Nor should the *Standards* be seen as requiring a specific curriculum. A curriculum is the way content is organized and presented in the classroom. The content embodied in the *Standards* can be organized and presented with different emphases and perspectives in many different curricula.

Evolution and the Nature of Science in the *National Science Education Standards*

Evolution and the nature of science are major topics in the content standards. The first mention of evolution is in the initial content standard, entitled "Unifying Concepts and Processes." This standard points out that conceptual and procedural schemes unify science disciplines and provide students with powerful ideas to help them understand the natural world. It is the only standard that extends across all grades, because the understanding and abilities associated with this standard need to be developed over an entire education.

The standard is as follows:

As a result of activities in grades K–12, all students should develop understanding and abilities aligned with the following concepts and processes:

- *Systems, order, and organization*
- *Evidence, models, and explanation*
- *Constancy, change, and measurement*
- ***Evolution** and equilibrium*
- *Form and function*

The guidance offered for the standard is to establish a broad context for thinking about evolution:

Evolution is a series of changes, some gradual and some sporadic, that accounts for the present form and function of objects, organisms, and natural and designed systems. The general idea of evolution is that the present arises from materials and forms of the past. Although evolution is most commonly associated with the biological theory explaining the process of descent with modification of organisms from common ancestors, evolution also describes changes in the universe.

With this unifying standard as a basis, the remaining content standards are organized by age group and discipline.

Grades K–4

The life science standard for grades K–4 is organized into the categories of characteristics of organisms, life cycles of organisms, and organisms and their environments. Evolution is not explicitly mentioned in these standards, but the text explains the basic things in life science that elementary school children ought to be able to understand and do:

During the elementary grades, children build understanding of biological concepts through direct experience with living things, their life cycles, and their habitats. These experiences emerge from the sense of wonder and natural interests of children who ask questions such as: "How do plants get food? How many different animals are there? Why do some animals eat other animals? What is the largest plant? Where did the dinosaurs go?" An understanding of the characteristics of organisms, life cycles of organisms, and of the complex interactions among all components of the natural environment begins with questions such as these and an understanding of how individual organisms maintain and continue life.

The intention of the K–4 standard is to develop the knowledge base that will be needed when the fundamental concepts of evolution are introduced in the middle and high school years.

Grades 5–8

For grades 5–8, the life science standard is the following:

As a result of their activities in grades 5–8, all students should develop understanding of:

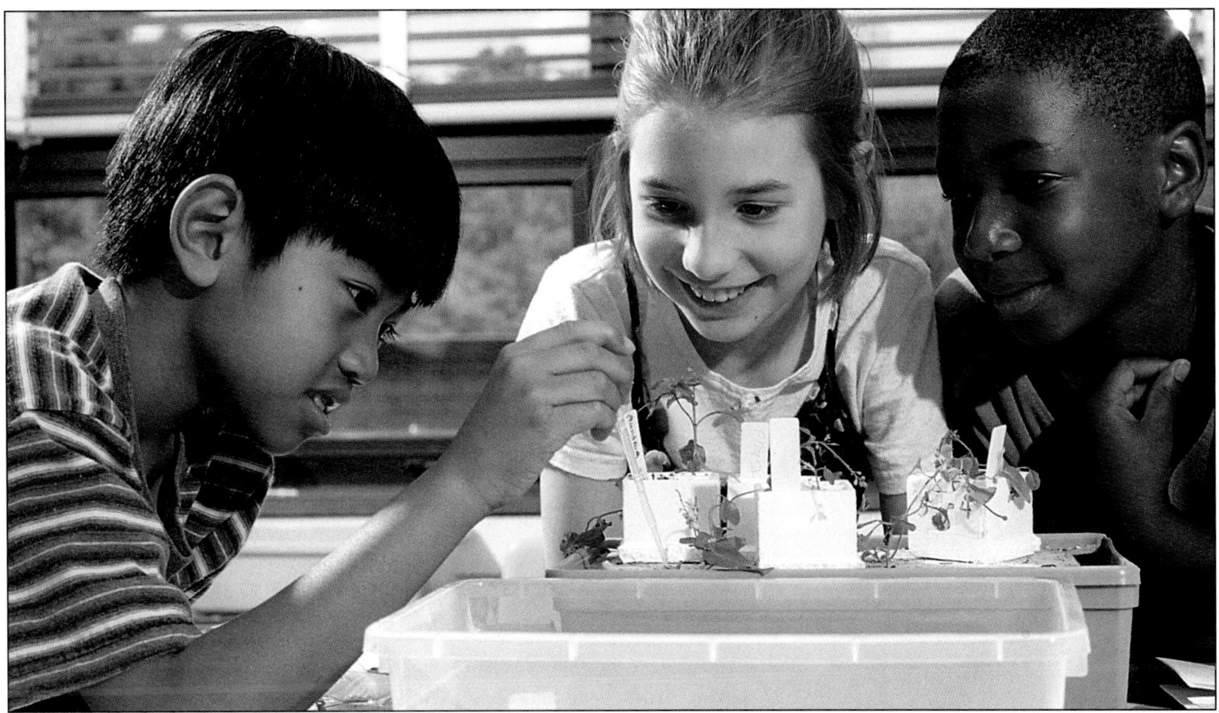

- *Structure and function in living systems*
- *Reproduction and heredity*
- ***Regulation and behavior***
- *Populations and ecosystems*
- ***Diversity and adaptations of organisms***

The guidance for this standard defines **regulation and behavior** as follows:

An organism's behavior evolves through adaptation to its environment. How a species moves, obtains food, reproduces, and responds to danger are based in the species' evolutionary history.

The text discusses **diversity and adaptations** as follows:

Diversity and Adaptations of Organisms

Millions of species of animals, plants, and microorganisms are alive today. Although different species might look dissimilar, the unity among organisms becomes apparent from an analysis of internal structures, the similarity of their chemical processes, and the evidence of common ancestry.

Biological evolution accounts for the diversity of species developed through gradual processes over many generations. Species acquire many of their unique characteristics through biological adaptation, which involves the selection of naturally occurring variations in populations. Biological adaptations include changes in structures, behaviors, or physiology that enhance survival and reproductive success in a particular environment.

Extinction of a species occurs when the environment changes and the adaptive characteristics of a species are insufficient to allow its survival. Fossils indicate that many organisms that lived long ago are extinct. Extinction of species is common; most of the species that have lived on the earth no longer exist.

The text accompanying the standard also discusses some of the difficulties encountered in teaching about adaptation:

Understanding adaptation can be particularly troublesome at this level. Many students think adaptation means that individuals change in major ways in response to environmental changes (that is, if the environment changes, individual organisms deliberately adapt).

In fact, as described in Chapter 2 of this book, adaptation occurs through natural selection, a topic described under the life science standards for grades 9–12.

The content standards also treat evolution in grades 5–8 in the section on earth's history. The standard reads as follows:

As a result of their activities in grades 5–8, all students should develop an understanding of:

- *Structure of the earth system*
- ***Earth's history***
- *Earth in the solar system*

The text discusses the importance of teaching students about earth systems and their interactions.

A major goal of science in the middle grades is for students to develop an understanding of earth and the solar system as a set of closely coupled systems. The idea of systems provides a framework in which students can investigate the four major interacting components of the earth system—geosphere (crust, mantle, and core), hydrosphere (water), atmosphere (air), and the biosphere (the realm of all living things). In this holistic approach to studying the planet, physical, chemical, and biological processes act within and among the four components on a wide range of time scales to change continuously earth's crust, oceans, atmosphere, and living organisms. Their study of earth's history provides students with some evidence about co-evolution of the planet's main features—the distribution of land and sea, features of the crust, the composition of the atmosphere, global climate, and populations of living organisms in the biosphere.

The material offering guidance for the standard explicitly ties the earth's history to the history of life:

Earth's History

The earth processes we see today, including erosion, movement of lithospheric plates, and changes in atmospheric composition, are similar to those that occurred in the past. Earth's history is also influenced by occasional catastrophes, such as the impact of an asteroid or comet.

Fossils provide important evidence of how life and environmental conditions have changed.

The standards for grades 5–8 cover the nature of science in the section on the history and nature of science:

As a result of activities in grades 5–8, all students should develop an understanding of:

- *Science as a human endeavor*
- ***Nature of science***
- ***History of science***

The guidance accompanying this standard offers the following discussion of these issues:

Nature of Science

Scientists formulate and test their explanations of nature using observation, experiments, and theoretical and mathematical models. Although all scientific ideas are tentative and subject to change and improvement in principle, for most major ideas in science, there is much experimental and observational confirmation. Those ideas are not likely to change greatly in the future. Scientists do and have changed their ideas about nature when they encounter new experimental evidence that does not match their existing explanations.

In areas where active research is being pursued and in which there is not a great deal of experimental or observational evidence and understanding, it is normal for scientists to differ with one another about the interpretation of the evidence or theory being considered. Different scientists might publish conflicting experimental results or might draw different conclusions from the same data. Ideally, scientists acknowledge such conflict and work towards finding evidence that will resolve their disagreement.

It is part of scientific inquiry to evaluate the results of scientific investigations, experiments, observations, theoretical models, and the explanations proposed by other scientists. Evaluation includes reviewing the experimental procedures, examining the evidence, identifying faulty reasoning, pointing out statements that go beyond the evidence, and suggesting alternative explanations for the same observations. Although scientists may disagree about explanations of phenomena, about interpretations of data, or about the value of rival theories, they do agree that questioning, response to criticism, and open communication are integral to the process of science. As scientific knowledge evolves, major disagreements are eventually resolved through such interactions between scientists.

History of Science

Many individuals have contributed to the traditions of science. Studying some of these individuals provides further understanding of scientific inquiry, science as a human endeavor, the nature of science, and the relationships between science and society.

In historical perspective, science has been practiced by different individuals in different cultures. In looking at the history of many peoples, one finds that scientists and engineers of high achievement are considered to be among the most valued contributors to their culture.

Tracing the history of science can show how difficult it was for scientific innovators to break through the accepted ideas of their time to reach the conclusions that we currently take for granted.

Grades 9–12

The life science standard for grades 9–12 directly addresses biological evolution. The standard reads as follows:

As a result of their activities in grades 9–12, all students should develop an understanding of:

- *The cell*
- *Molecular basis of heredity*
- ***Biological evolution***
- *Interdependence of organisms*
- *Matter, energy, and organization in living systems*
- *Behavior of organisms*

The guidance for the life science standard describes the major themes of evolutionary theory:

Biological Evolution

Species evolve over time. Evolution is the consequence of the interactions of (1) the potential for a species to increase its numbers, (2) the genetic variability of offspring due to mutation and recombination of genes, (3) a finite supply of the resources required for life, and (4) the ensuing selection by the environment of those offspring better able to survive and leave offspring.

The great diversity of organisms is the result of more than 3.5 billion years of evolution that has filled every available niche with life forms.

Natural selection and its evolutionary consequences provide a scientific explanation for the fossil record of ancient life forms, as well as for the striking molecular similarities observed among the diverse species of living organisms.

The millions of different species of plants, animals, and microorganisms that live on earth today are related by descent from common ancestors.

Biological classifications are based on how organisms are related. Organisms are classified into a hierarchy of groups and subgroups based on similarities which reflect their evolutionary relationships. Species is the most fundamental unit of classification.

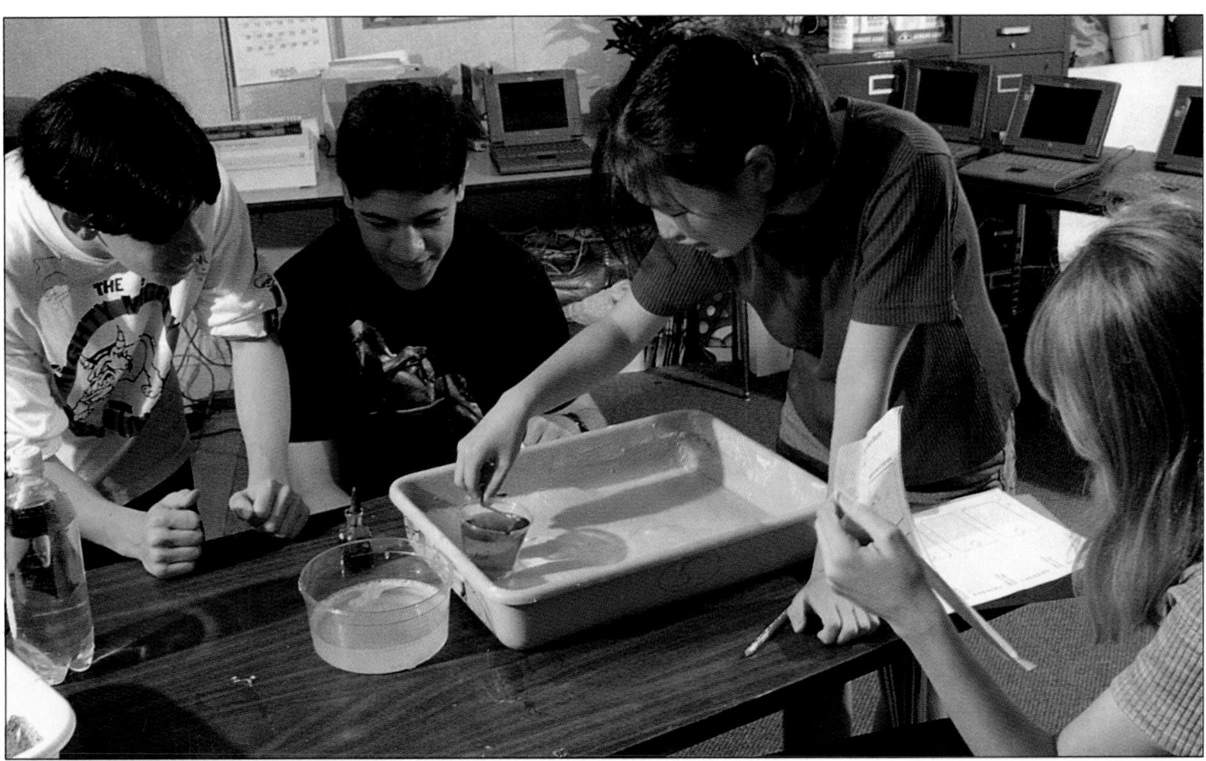

The text following the standard describes some of the difficulties that students can have in comprehending the basic concepts of evolution.

Students have difficulty with the fundamental concepts of evolution. For example, students often do not understand natural selection because they fail to make a conceptual connection between the occurrence of new variations in a population and the potential effect of those variations on the long-term survival of the species. One misconception that teachers may encounter involves students attributing new variations to an organism's need, environmental conditions, or use. With some help, students can understand that, in general, mutations occur randomly and are selected because they help some organisms survive and produce more offspring. Other misconceptions center on a lack of understanding of how a population changes as a result of differential reproduction (some individuals producing more offspring), as opposed to all individuals in a population changing. Many misconceptions about the process of natural selection can be changed through instruction.

Finally, evolution is discussed again in the guidance following the earth and space science standard:

As a result of their activities in grades 9–12, all students should develop an understanding of:

- *Energy in the earth system*
- *Geochemical cycles*
- ***Origin and evolution of the earth system***
- ***Origin and evolution of the universe***

The discussions of the origin and evolution of the earth system and the universe relate evolution to universal physical processes:

The Origin and Evolution of the Earth System

The sun, the earth, and the rest of the solar system formed from a nebular cloud of dust and gas 4.5 billion years ago. The early earth was very different from the planet we live on today.

Geologic time can be estimated by observing rock sequences and using fossils to correlate the sequences at various locations. Current methods include using the known decay rates of radioactive isotopes present in rocks to measure the time since the rock was formed.

Interactions among the solid earth, the oceans, the atmosphere, and organisms have resulted in the ongoing evolution of the earth system. We can observe some changes such as earthquakes and volcanic eruptions on a human time scale, but many processes such as mountain building and plate movements take place over hundreds of millions of years.

Evidence for one-celled forms of life—the bacteria—extends back more than 3.5 billion years. The evolution of life caused dramatic changes in the composition of the earth's atmosphere, which did not originally contain oxygen.

The Origin and Evolution of the Universe

The origin of the universe remains one of the greatest questions in science. The "big bang" theory places the origin between 10 and 20 billion years ago, when the universe began in a hot dense state; according to this theory, the universe has been expanding ever since.

Early in the history of the universe, matter, primarily the light atoms hydrogen and helium, clumped together by gravitational attraction to form countless trillions of stars. Billions of galaxies, each of which is a gravitationally bound cluster of billions of stars, now form most of the visible mass in the universe.

Stars produce energy from nuclear reactions, primarily the fusion of hydrogen to form helium. These and other processes in stars have led to the formation of all the other elements.

The standard for the history and nature of science elaborates on the knowledge established in previous years:

As a result of activities in grades 9–12, all students should develop an understanding of:

- *Science as a human endeavor*
- ***Nature of scientific knowledge***
- ***Historical perspectives***

The discussion of this standard relates the nature of science explicitly to many of the problems that arise in the teaching of evolution.

Nature of Scientific Knowledge

Science distinguishes itself from other ways of

knowing and from other bodies of knowledge through the use of empirical standards, logical arguments, and skepticism, as scientists strive for the best possible explanations about the natural world.

Scientific explanations must meet certain criteria. First and foremost, they must be consistent with experimental and observational evidence about nature, and must make accurate predictions, when appropriate, about systems being studied. They should also be logical, respect the rules of evidence, be open to criticism, report methods and procedures, and make knowledge public. Explanations on how the natural world changes based on myths, personal beliefs, religious values, mystical inspiration, superstition, or authority may be personally useful and socially relevant, but they are not scientific.

Because all scientific ideas depend on experimental and observational confirmation, all scientific knowledge is, in principle, subject to change as new evidence becomes available. The core ideas of science such as the conservation of energy or the laws of motion have been subjected to a wide variety of confirmations and are therefore unlikely to change in the areas in which they have been tested. In areas where data or understanding are incomplete, such as the details of human evolution or questions surrounding global warming, new data may well lead to changes in current ideas or resolve current conflicts. In situations where information is still fragmentary, it is normal for scientific ideas to be incomplete, but this is also where the opportunity for making advances may be greatest.

Historical Perspectives

In history, diverse cultures have contributed scientific knowledge and technologic inventions. Modern science began to evolve rapidly in Europe several hundred years ago. During the past two centuries, it has contributed significantly to the industrialization of Western and non-Western cultures. However, other, non-European cultures have developed scientific ideas and solved human problems through technology.

Usually, changes in science occur as small modifications in extant knowledge. The daily work of science and engineering results in incremental advances in our understanding of the world and our ability to meet human needs and aspirations. Much can be learned about the internal workings of science and the nature of science from study of individual scientists, their daily work, and their efforts to advance scientific knowledge in their area of study.

Conclusion

The material addressing evolution in the *National Science Education Standards* is embedded within the full range of content standards describing what students should know, understand, and be able to do in the natural sciences. Used in conjunction with standards for other parts of the science education system, the content standards—and their treatment of evolution—point toward the levels of scientific literacy needed to meet the challenges of the twenty-first century.

NOTES

1. National Research Council. 1996. *National Science Education Standards*. Washington, DC: National Academy Press. www.nap.edu/readingroom/books/nses
2. American Association for the Advancement of Science. 1993. *Benchmarks for Science Literacy*. Project 2061. New York: Oxford University Press. www.aaas.org
3. National Science Teachers Association. 1993. *Scope, Sequence, and Coordination of Secondary School Science. Vol. 1. The Content Core: A Guide for Curriculum Designers*. rev. ed. Arlington, VA: NSTA. www.nsta.org

5

Frequently Asked Questions About Evolution and the Nature of Science

Teachers often face difficult questions about evolution, many from parents and others who object to evolution being taught. Science has good answers to these questions, answers that draw on the evidence supporting evolution and on the nature of science. This chapter presents short answers to some of the most commonly asked questions.

Definitions

What is evolution?

Evolution in the broadest sense explains that what we see today is different from what existed in the past. Galaxies, stars, the solar system, and earth have changed through time, and so has life on earth.

Biological evolution concerns changes in living things during the history of life on earth. It explains that living things share common ancestors. Over time, evolutionary change gives rise to new species. Darwin called this process "descent with modification," and it remains a good definition of biological evolution today.

What is "creation science"?

The ideas of "creation science" derive from the conviction that God created the universe—including humans and other living things—all at once in the relatively recent past. However, scientists from many fields have examined these ideas and have found them to be scientifically insupportable. For example, evidence for a very young earth is incompatible with many different methods of establishing the age of rocks. Furthermore, because the basic proposals of creation science are not subject to test and verification, these ideas do not meet the criteria for science. Indeed, U.S. courts have ruled that ideas of creation science are religious views and cannot be taught when evolution is taught.

The Supporting Evidence

How can evolution be scientific when no one was there to see it happen?

This question reflects a narrow view of how science works. Things in science can be studied even if they cannot be directly observed or experimented on. Archaeologists study past cultures by examining the artifacts those cultures left behind. Geologists can describe past changes in sea level by studying the marks ocean waves left on rocks. Paleontologists study the fossilized remains of organisms that lived long ago.

Something that happened in the past is thus not "off limits" for scientific study. Hypotheses can be made about such phenomena, and these hypotheses can be tested and can lead to solid conclusions. Furthermore, many key aspects of evolution occur in relatively short periods that can be observed directly—such as the evolution in bacteria of resistance to antibiotics.

Isn't evolution just an inference?

No one saw the evolution of one-toed horses from three-toed horses, but that does not mean that we cannot be confident that horses evolved. Science is practiced in many ways besides direct observation and experimentation. Much scientific discovery is done through indirect experimentation and observation in which inferences are made, and hypotheses generated from those inferences are tested.

For instance, particle physicists cannot directly observe subatomic particles because the particles are too small. They must make inferences about the weight, speed, and other properties of the particles based on other observations. A logical hypothesis might be something like this: If the weight of this particle is *Y*, when I bombard it, *X* will happen. If *X* does not happen, then the hypothesis is disproved. Thus, we can learn about the natural world even if we cannot directly observe a phenomenon —and that is true about the past, too.

In historical sciences like astronomy, geology, evolutionary biology, and archaeology, logical inferences are made and then tested against data. Sometimes the test cannot be made until new data are available, but a great deal has been done to help us understand the past. For example, scorpionflies (*Mecoptera*) and true flies (*Diptera*) have enough similarities that entomologists consider them to be closely related. Scorpionflies have four wings of about the same size, and true flies have a large front pair of wings but the back pair is replaced by small club-shaped structures. If *Diptera* evolved from *Mecoptera*, as comparative anatomy suggests, scientists predicted that a fossil fly with four wings might be found—and in 1976 this is exactly what was discovered. Furthermore, geneticists have found that the number of wings in flies can be changed through mutations in a single gene.

Evolution is a well-supported theory drawn from a variety of sources of data, including observations about the fossil record, genetic information, the distribution of plants and animals, and the similarities across species of anatomy and development. Scientists have inferred that descent with modification offers the best scientific explanation for these observations.

Is evolution a fact or a theory?

The theory of evolution explains how life on earth has changed. In scientific terms, "theory" does not mean "guess" or "hunch" as it does in everyday usage. Scientific theories are explanations of natural phenomena built up logically from testable observations and hypotheses. Biological evolution is the best scientific explanation we have for the enormous range of observations about the living world.

Scientists most often use the word "fact" to describe an observation. But scientists can also use fact to mean something that has been tested or observed so many times that there is no longer a compelling reason to keep testing or looking for examples. The occurrence of evolution in this sense is a fact. Scientists no longer question whether descent with modification occurred because the evidence supporting the idea is so strong.

Why isn't evolution called a law?

Laws are generalizations that *describe* phenomena, whereas theories *explain* phenomena. For example, the laws of thermodynamics describe what will happen under certain circumstances; thermodynamics theories explain why these events occur.

Laws, like facts and theories, can change with better data. But theories do not develop into laws with the accumulation of evidence. Rather, theories are the goal of science.

Don't many famous scientists reject evolution?

No. The scientific consensus around evolution is overwhelming. Those opposed to the teaching of evolution sometimes use quotations from prominent scientists out of context to claim that scientists do not support evolution. However, examination of the quotations reveals that the scientists are actually disputing some aspect of *how* evolution occurs, not *whether* evolution occurred. For example, the biologist Stephen Jay Gould once wrote that "the extreme rarity of transitional forms in the fossil record persists as the trade secret of paleontology." But Gould, an accomplished paleontologist and eloquent educator about evolution, was arguing about *how* evolution takes place. He was discussing whether the rate of change of species is constant and gradual or whether it takes place in bursts after long periods when little change occurs—an idea known as punctuated equilibrium. As Gould writes in response, "This quotation, although accurate as a partial citation, is dishonest in leaving out the following explanatory material showing my true purpose—to discuss rates of evolutionary change, not to deny the fact of evolution itself."

Gould defines punctuated equilibrium as follows:

Punctuated equilibrium is neither a creationist idea nor even a non-Darwinian evolutionary theory about sudden change that produces a new species all at once in a single generation. Punctuated equilibrium accepts the conventional idea that new species form over hundreds or thousands of generations and through an extensive series of intermediate stages. But geological time is so long that even a few thousand years may appear as a mere "moment" relative to the several million years of existence for most species. Thus, rates of evolution vary enormously and new species may appear to arise "suddenly" in geological time, even though the time involved would seem long, and the change very slow, when compared to a human lifetime.

Isn't the fossil record full of gaps?

Though significant gaps existed in the fossil record in the 19th century, many have been filled in. In addition, the consistent pattern of ancient to modern species found in the fossil record is strong evidence for evolution. The plants and animals living today are not like the plants and animals of the remote past. For example, dinosaurs were extinct long before humans walked the earth. We know this because no human remains have ever been found in rocks dated to the dinosaur era.

Some changes in populations might occur too rapidly to leave many transitional fossils. Also, many organisms were very unlikely to leave fossils, either because of their habitats or because they had no body parts that could easily be fossilized. However, in many cases, such as between primitive fish and amphibians, amphibians and reptiles, reptiles and mammals, and reptiles and birds, there are excellent transitional fossils.

Can evolution account for new species?

One argument sometimes made by supporters of "creation science" is that natural selection can produce minor changes within species, such as changes in color or beak size, but cannot generate new species from pre-existing species. However, evolutionary biologists have documented many cases in which new species have appeared in recent years (some of these cases are discussed in Chapter 2). Among most plants and animals, speciation is an extended process, and a single human observer can witness only a part of this process. Yet these observations of evolution at work provide powerful confirmation that evolution forms new species.

If humans evolved from apes, why are there still apes?

Humans did not evolve from modern apes, but humans and modern apes shared a common ancestor, a species that no longer exists. Because we shared a recent common ancestor with chimpanzees and gorillas, we have many anatomical, genetic, biochemical, and even behavioral similarities with the African great apes. We are less similar to the Asian apes—orangutans and gibbons—and even less similar to monkeys, because we shared common ancestors with these groups in the more distant past.

Evolution is a branching or splitting process in which populations split off from one another and gradually become different. As the two groups become isolated from each other, they stop sharing genes, and eventually genetic differences increase until members of the groups can no longer interbreed. At this point, they have become separate species. Through time, these two species might give rise to new species, and so on through millennia.

Doesn't the sudden appearance of all the "modern groups" of animals during the Cambrian explosion prove creationism?

During the Cambrian explosion, primitive representatives of the major phyla of invertebrate animals appeared—hard-shelled organisms like mollusks and arthropods. More modern representatives of these invertebrates appeared gradually through the Cambrian and the Ordovician periods. "Modern groups" like terrestrial vertebrates and flowering plants were not present. It is not true that "all the modern groups of animals" appeared during this period.

Also, Cambrian fossils did not appear spontaneously. They had ancestors in the Precambrian period, but because these Precambrian forms were soft-bodied, they left fewer fossils. A characteristic of the Cambrian fossils is the evolution of hard body parts, which greatly improved the chance of fossilization. And even without fossils, we can infer relationships among organisms from biochemical information.

Religious Issues

Can a person believe in God and still accept evolution?

Many do. Most religions of the world do not have any direct conflict with the idea of evolution. Within the Judeo-Christian religions, many people believe that God works through the process of evolution. That is, God has created both a world that is ever-changing and a mechanism through which creatures can adapt to environmental change over time.

At the root of the apparent conflict between some religions and evolution is a misunderstanding of the critical difference between religious and scientific ways of knowing. Religions and science answer different questions about the world. Whether there is a purpose to the universe or a purpose for human existence are not questions for science. Religious and scientific ways of knowing have played, and will continue to play, significant roles in human history.

No one way of knowing can provide all of the answers to the questions that humans ask. Consequently, many people, including many scientists, hold strong religious beliefs and simultaneously accept the occurrence of evolution.

Aren't scientific beliefs based on faith as well?

Usually "faith" refers to beliefs that are accepted without empirical evidence. Most religions have tenets of faith. Science differs from religion because it is the nature of science to test and retest explanations against the natural world. Thus, scientific explanations are likely to be built on and modified with new information and new ways of looking at old information. This is quite different from most religious beliefs.

Therefore, "belief" is not really an appropriate term to use in science, because testing is such an important part of this way of knowing. If there is a component of faith to science, it is the assumption that the universe operates according to regularities—for example, that the speed of light will not change tomorrow. Even the assumption of that regularity is often tested—and thus far has held up well. This "faith" is very different from religious faith.

Science is a way of knowing about the natural world. It is limited to explaining the natural world through natural causes. Science can say nothing about the supernatural. Whether God exists or not is a question about which science is neutral.

Legal Issues

Why can't we teach creation science in my school?

The courts have ruled that "creation science" is actually a religious view. Because public schools must be religiously neutral under the U.S. Constitution, the courts have held that it is unconstitutional to present creation science as legitimate scholarship.

In particular, in a trial in which supporters of creation science testified in support of their view, a district court declared that creation science does not meet the tenets of science as scientists use the term (*McLean v. Arkansas Board of Education*). The Supreme Court has held that it is illegal to require that creation science be taught when evolution is taught (*Edwards v. Aguillard*). In addition, district courts have decided that individual teachers cannot advocate creation science on their own (*Peloza v. San Juan Capistrano School District* and *Webster v. New Lennox School District*).

Teachers' organizations such as the National Science Teachers Association, the National Association of Biology Teachers, the National Science Education Leadership Association, and many others also have rejected the science and pedagogy of creation science and have strongly discouraged its presentation in the public schools. (Statements from some of these organizations appear in Appendix C.) In addition, a coalition of religious and other organizations has noted in "A Joint Statement of Current Law" (see Appendix B) that "in science class, [schools] may present only genuinely scientific critiques of, or evidence for, any explanation of life on earth, but not religious critiques (beliefs unverifiable by scientific methodology)."

Some argue that "fairness" demands the teaching of creationism along with evolution. But a science curriculum should cover science, not the religious views of particular groups or individuals.

Educational Issues

If evolution is taught in schools, shouldn't creationism be given equal time?

Some religious groups deny that microorganisms cause disease, but the science curriculum should not therefore be altered to reflect this belief. Most people agree that students should be exposed to the best possible scholarship in each field. That scholarship is evaluated by professionals and educators in those fields. In science, scientists as well as educators have concluded that evolution—and only evolution—should be taught in science classes because it is the only *scientific* explanation for why the universe is the way it is today.

Many people say that they want their children to be exposed to creationism in school, but there are thousands of different ideas about creation among the world's people. Comparative religions might comprise a worthwhile field of study but not one appropriate for a science class. Furthermore, the U.S. Constitution states that schools must be religiously neutral, so legally a teacher could not present any particular creationist view as being more "true" than others.

Why should teachers teach evolution when they already have so many things to teach and can cover biology without mentioning evolution?

Teachers face difficult choices in deciding what to teach in their limited time, but some ideas are of central importance in each discipline. In biology, evolution is such an idea. Biology is sometimes taught as a list of facts, but if evolution is introduced early in a class and in an uncomplicated manner, it can tie many disparate facts together. Most important, it offers a way to understand the astonishing complexity, diversity, and activity of the modern world. Why are there so many different types of organisms? What is the response of a species or community to a changing environment? Why is it so difficult to develop antibiotics and insecticides that are useful for more than a decade or two? All of these questions are easily discussed in terms of evolution but are difficult to answer otherwise.

A lack of instruction about evolution also can hamper students when they need that information to take other classes, apply for college or medical school, or make decisions that require a knowledge of evolution.

Should students be given lower grades for not believing in evolution?

No. Children's personal views should have no effect on their grades. Students are not under a compulsion to accept evolution. A grade reflects a teacher's assessment of a student's understanding. If a child does not understand the basic ideas of evolution, a grade could and should reflect that lack of understanding, because it is quite possible to comprehend things that are not believed.

Can evolution be taught in an inquiry-based fashion?

Any science topic can be taught in an inquiry-oriented manner, and evolution is particularly amenable to this approach. At the core of inquiry-oriented instruction is the provision for students to collect data (or be given data when collection is not possible) and to analyze the data to derive patterns, conclusions, and hypotheses, rather than just learning facts. Students can use many data sets from evolution (such as diagrams of anatomical differences in organisms) to derive patterns or draw connections between morphological forms and environmental conditions. They then can use their data sets to test their hypotheses.

Students also can collect data in real time. For example, they can complete extended projects involving crossbreeding of fruit flies or plants to illustrate the genetic patterns of inheritance and the influence of the environment on survival. In this way, students can develop an understanding of evolution, scientific inquiry, and the nature of science.

Manatee

6

Activities for Teaching About Evolution and the Nature of Science

Prior chapters in this volume answer the what and why questions of teaching about evolution and the nature of science. As every educator knows, such discussions only set a stage. The actual play occurs when science teachers act on the basic content and well-reasoned arguments for inclusion of evolution and the nature of science in school science programs.

This chapter goes beyond discussions of content and rationales. It presents, as examples of investigative teaching exercises, eight activities that science teachers can use as they begin developing students' understandings and abilities of evolution and the nature of science. The following descriptions briefly introduce each activity.

ACTIVITY 1: Introducing Inquiry and the Nature of Science

This activity introduces basic procedures involved in inquiry and concepts describing the nature of science. In the first portion of the activity the teacher uses a numbered cube to involve students in asking a question—what is on the unseen bottom of the cube?—and the students propose an explanation based on their observations. Then the teacher presents the students with a second cube and asks them to use the available evidence to propose an explanation for what is on the bottom of this cube. Finally, students design a cube that they exchange and use for an evaluation. This activity provides students with opportunities to learn the abilities and understandings aligned with science as inquiry and the nature of science as described in the *National Science Education Standards*.[1] Designed for grades 5 through 12, the activity requires a total of four class periods to complete. Lower grade levels might only complete the first cube and the evaluation where students design a problem based on the cube activity.

ACTIVITY 2: The Formulation of Explanations: An Invitation to Inquiry on Natural Selection

This activity uses the concept of natural selection to introduce the idea of formulating and testing a scientific hypothesis. Through a focused discussion approach, the teacher provides information and allows students time to think, interact with peers, and propose explanations for observations described by the teacher. The teacher then provides more information, and the students continue their discussion based on the new information. This activity will help students in grades 5 through 8 develop abilities related to scientific inquiry and formulate understandings about the nature of science.

ACTIVITY 3: Investigating Natural Selection

In this activity, the students investigate one mechanism for evolution through a simulation that models the principles of natural selection and helps answer the question: How might biological change have occurred and been reinforced over time? The activity is designed for grades 9 through 12 and requires three class periods.

ACTIVITY 4: Investigating Common Descent: Formulating Explanations and Models

In this activity, students formulate explanations and models that simulate structural and biochemical data as they investigate the misconception that humans evolved from apes. The investigations require two 45-minute periods. They are designed for use in grades 9 through 12.

ACTIVITY 5: Proposing Explanations for Fossil Footprints

In this investigation, students observe and interpret "fossil footprint" evidence. From the evidence, they are asked to construct defensible hypotheses or explanations for events that took place in the geologic past. Estimated time requirements for this activity: two class periods. This activity is designed for grades 5 through 8.

ACTIVITY 6: Understanding Earth's Changes Over Time

Comparing the magnitude of geologic time to spans of time within a person's own lifetime is difficult for many students. In this activity, students use a long paper strip and a reasonable scale to represent visually all of geologic time, including significant events in the development of life on earth as well as recent human events. The investigation requires two class periods and is appropriate for grades 5 through 12.

ACTIVITY 7: Proposing the Theory of Biological Evolution: Historical Perspective

This activity uses historical perspectives and the theme of evolution to introduce students to the nature of science. The teacher has students read short excerpts of original statements on evolution from Jean Lamarck, Charles Darwin, and Alfred Russel Wallace. These activities are intended as either supplements to other investigations or core activities. Designed for grades 9 through 12, the activities should be used as part of three class periods.

ACTIVITY 8: Connecting Population Growth and Biological Evolution

In this activity, students develop a model of the mathematical nature of population growth. The investigation provides an excellent opportunity for consideration of population growth of plant and animal species and the relationship to mechanisms promoting natural selection. This activity will require two class periods and is appropriate for grades 5 through 12.

The activities in this chapter do not represent a curriculum. They are directed, instead, toward other purposes.

First, they present examples of standards-based instructional materials. In this case, the level of organization is an activity—one to five days of lessons—and not a larger level of organization such as a unit of several weeks, a semester, or a year. Also, these exercises generally do not use biological materials, such as fruit flies, or computer simulations. The use of these instructional materials in the curriculum greatly expands the range of possible investigations.

Second, these activities demonstrate how existing exercises can be recast to emphasize the importance of inquiry and the fundamental concepts of evolution. Each of these exercises was derived from already existing activities that were revised to reflect the *National Science Education Standards*. For each exercise, student outcomes drawn from the *Standards* are listed to focus attention on the concepts and abilities that students are meant to develop.

Third, the activities demonstrate some, but not all, of the criteria for curricula to be described in Chapter 7. For example, several of the activities emphasize inquiry and the nature of science while others focus on concepts related to evolution. All activities use an instructional model, described in the next section, that increases coherence and enhances learning.

Finally, there remains a paucity of instructional materials for teaching evolution and the nature of science. Science teachers who recognize this need are encouraged to develop new materials and lessons to introduce the themes of evolution and the nature of science. (See http://www4.nas.edu/opus/evolve.nsf)

Developing Students' Understanding and Abilities: The Curriculum Perspective

For students to develop an understanding of evolution and the nature of science requires many years and a variety of educational experiences.

Teachers cannot rely on single lessons, chapters, or biology and earth science courses for students to integrate the ideas presented in this document into their own understanding. In early grades (K–4) students might learn the fundamental concepts associated with "characteristics of organisms," "life cycles," and "organisms and environments." In middle grades they learn more about "reproduction and heredity" and "diversity and adaptation of organisms." Such learning experiences, as described in the *National Science Education Standards*, set a firm foundation for the study of biological evolution in grades 9–12.

The slow and steady development of concepts such as evolution and related ideas such as natural selection and common descent requires careful consideration of the overall structure and sequence of learning experiences. Although this chapter does not propose a curriculum or a curriculum framework, current efforts by Project 2061 of the American Association for the Advancement of Science (AAAS) demonstrate the interrelated nature of students' understanding of science concepts and emphasize the importance of well-designed curricula at several levels of organization (for example, activities, units, and school science programs). The figure on the next page presents the "Growth-of-Understanding Map for Evolution and Natural Selection" based on *Benchmarks for Science Literacy*.[2]

Developing Student Understanding and Abilities: The Instructional Perspective

The activities in the chapter incorporate an instructional model, summarized in the accompanying box, that includes five steps: engagement, exploration, explanation, elaboration, and evaluation. Just as scientific investigations originate with a question that engages a scientist, so too must students engage in the activities of learning. The activities therefore begin with a strategic question that gets students thinking about the content of the lesson.

Once engaged, students need time to explore ideas before concepts begin to make sense. In this exploration phase, students try their ideas, ask questions, and look for possible answers to questions. Students use inquiry strategies; they try to

An Instructional Model

ENGAGE This phase of the instructional model initiates the learning task. The activity should (1) make connections between past and present learning experiences and (2) anticipate activities and focus students' thinking on the learning outcomes of current activities. Students should become mentally engaged in the concept, process, or skill to be explored.

EXPLORE This phase of the teaching model provides students with a common base of experiences within which they identify and develop current concepts, processes, and skills. During this phase, students actively explore their environment or manipulate materials.

EXPLAIN This phase of the instructional model focuses students' attention on a particular aspect of their engagement and exploration experiences and provides opportunities for them to develop explanations and hypotheses. This phase also provides opportunities for teachers to introduce a formal label or definition for a concept, process, skill, or behavior.

ELABORATE This phase of the teaching model challenges and extends students' conceptual understanding and allows further opportunity for students to test hypotheses and practice desired skills and behaviors. Through new experiences, the students develop a deeper and broader understanding, acquire more information, and develop and refine skills.

EVALUATE This phase of the teaching model encourages students to assess their understanding and abilities and provides opportunities for teachers to evaluate student progress toward achieving the educational objectives.

EVOLUTION AND NATURAL SELECTION

This draft map shows the development of ideas, and relationships between them, that contribute to a key element of science literacy, understanding biological evolution. The boxes contain specific learning goals and include a code that refers to the corresponding Benchmark *or* Science for All Americans *passage.*

The arrows signify that one learning goal contributes to an understanding of another. Double-headed arrows imply mutual support. The gray box around three learning goals in the K-2 range shows that these goals are closely related and any sequencing is unimportant. (Arrows that attach to the outside of the gray box include the whole group.)

Often, ideas from a topic area not represented on this map play a role in understanding biological evolution. For example, an understanding of heredity would be required to understand the origin and passing on of new traits. Ideas from other fields may also contribute to understanding evolution, such as knowledge of isotopic dating techniques to account for the enormous amount of time that evolution theory encompasses.

This map is a work in progress intended for publication in the Atlas of Science Literacy, *AAAS—Project 2061.*

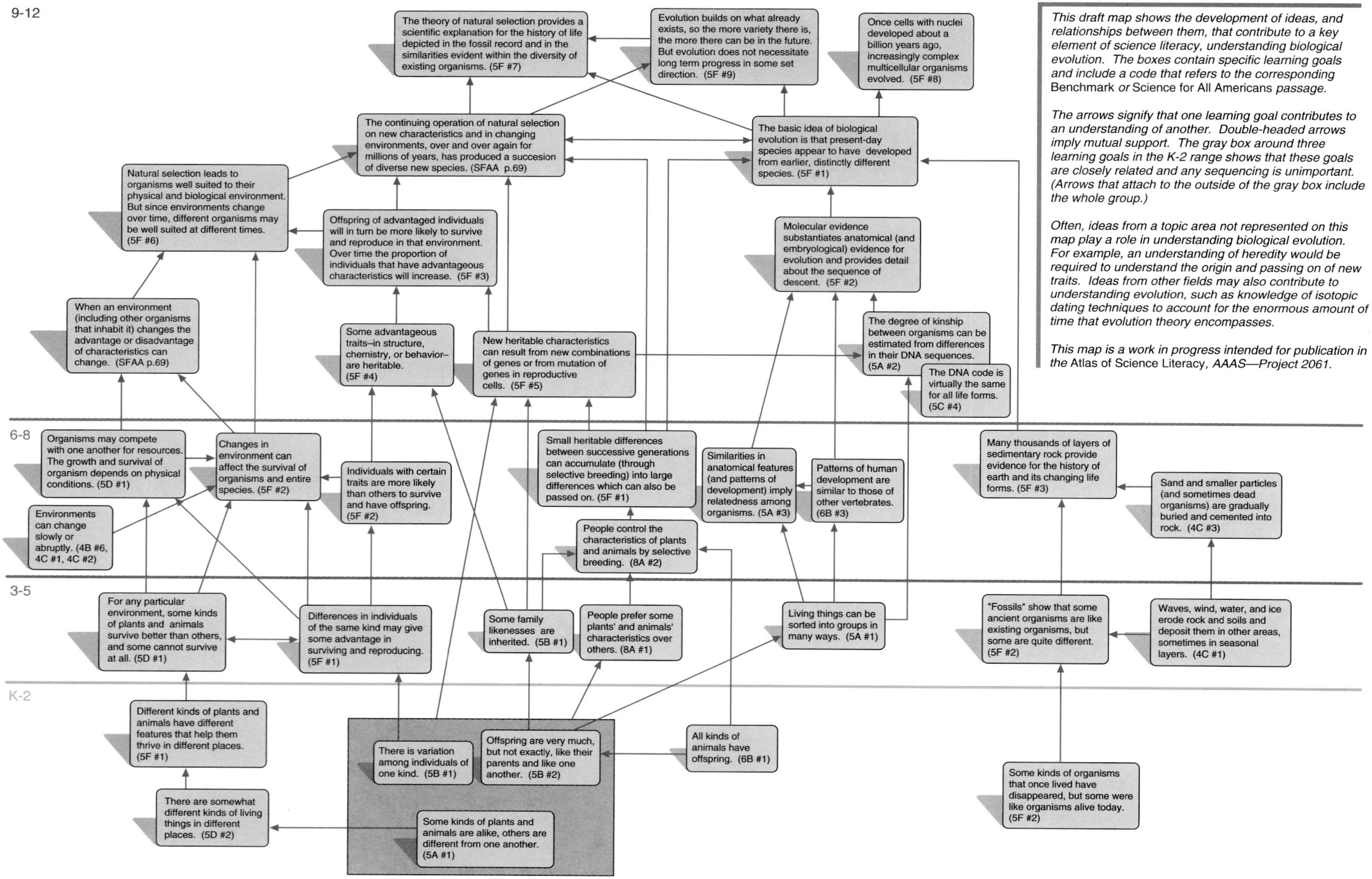

9-12

The theory of natural selection provides a scientific explanation for the history of life depicted in the fossil record and in the similarities evident within the diversity of existing organisms. (5F #7)

Evolution builds on what already exists, so the more variety there is, the more there can be in the future. But evolution does not necessitate long term progress in some set direction. (5F #9)

Once cells with nuclei developed about a billion years ago, increasingly complex multicellular organisms evolved. (5F #8)

The continuing operation of natural selection on new characteristics and in changing environments, over and over again for millions of years, has produced a succesion of diverse new species. (SFAA p.69)

The basic idea of biological evolution is that present-day species appear to have developed from earlier, distinctly different species. (5F #1)

Natural selection leads to organisms well suited to their physical and biological environment. But since environments change over time, different organisms may be well suited at different times. (5F #6)

Offspring of advantaged individuals will in turn be more likely to survive and reproduce in that environment. Over time the proportion of individuals that have advantageous characteristics will increase. (5F #3)

Molecular evidence substantiates anatomical (and embryological) evidence for evolution and provides detail about the sequence of descent. (5F #2)

When an environment (including other organisms that inhabit it) changes the advantage or disadvantage of characteristics can change. (SFAA p.69)

Some advantageous traits–in structure, chemistry, or behavior–are heritable. (5F #4)

New heritable characteristics can result from new combinations of genes or from mutation of genes in reproductive cells. (5F #5)

The degree of kinship between organisms can be estimated from differences in their DNA sequences. (5A #2)

The DNA code is virtually the same for all life forms. (5C #4)

6-8

Organisms may compete with one another for resources. The growth and survival of organism depends on physical conditions. (5D #1)

Changes in environment can affect the survival of organisms and entire species. (5F #2)

Individuals with certain traits are more likely than others to survive and have offspring. (5F #2)

Small heritable differences between successive generations can accumulate (through selective breeding) into large differences which can also be passed on. (5F #1)

Similarities in anatomical features (and patterns of development) imply relatedness among organisms. (5A #3)

Patterns of human development are similar to those of other vertebrates. (6B #3)

Many thousands of layers of sedimentary rock provide evidence for the history of earth and its changing life forms. (5F #3)

Environments can change slowly or abruptly. (4B #6, 4C #1, 4C #2)

People control the characteristics of plants and animals by selective breeding. (8A #2)

Sand and smaller particles (and sometimes dead organisms) are gradually buried and cemented into rock. (4C #3)

3-5

For any particular environment, some kinds of plants and animals survive better than others, and some cannot survive at all. (5D #1)

Differences in individuals of the same kind may give some advantage in surviving and reproducing. (5F #1)

Some family likenesses are inherited. (5B #1)

People prefer some plants' and animals' characteristics over others. (8A #1)

Living things can be sorted into groups in many ways. (5A #1)

"Fossils" show that some ancient organisms are like existing organisms, but some are quite different. (5F #2)

Waves, wind, water, and ice erode rock and soils and deposit them in other areas, sometimes in seasonal layers. (4C #1)

K-2

Different kinds of plants and animals have different features that help them thrive in different places. (5F #1)

There are somewhat different kinds of living things in different places. (5D #2)

There is variation among individuals of one kind. (5B #1)

Offspring are very much, but not exactly, like their parents and like one another. (5B #2)

Some kinds of plants and animals are alike, others are different from one another. (5A #1)

All kinds of animals have offspring. (6B #1)

Some kinds of organisms that once lived have disappeared, but some were like organisms alive today. (5F #2)

relate their ideas to those of other students and to what scientists already know about evolution.

In the third step, students can propose answers and develop hypotheses. Also in this step, the teacher explains what scientists know about the questions. This is the step when teachers should make the major concepts explicit and clear to the students.

Educators understand that informing students about a concept does not necessarily result in their immediate comprehension and understanding of the idea. These activities therefore provide a step referred to as elaboration in which students have opportunities to apply their ideas in new and slightly different situations.

Finally, how well do students understand the concepts, or how successful are they at applying the desired skills? These are the questions to be answered during the evaluation phase. Ideally, evaluations are more than tests. Students should have opportunities to see if their ideas can be applied in new situations and to compare their understanding with scientific explanations of the same phenomena.

ACTIVITY 1

Introducing Inquiry and the Nature of Science

This activity introduces basic procedures involved in inquiry and concepts describing the nature of science. In the first portion of the activity the teacher uses a numbered cube to involve students in asking a question—what is on the bottom?— and the students propose an explanation based on their observations. Then the teacher presents the students with a second cube and asks them to use the available evidence to propose an explanation for what is on the bottom of this cube. Finally, students design a cube that they exchange and use for an evaluation. This activity provides students with opportunities to learn the abilities and understandings aligned with science as inquiry and the nature of science as described in the *National Science Education Standards.* Designed for grades 5 through 12, the activity requires a total of four class periods to complete. Lower grade levels might only complete the first cube and the evaluation where students design a problem based on the cube activity.

Standards-Based Outcomes

This activity provides all students with opportunities to develop abilities of scientific inquiry as described in the *National Science Education Standards*. Specifically, it enables them to:

- identify questions that can be answered through scientific investigations,
- design and conduct a scientific investigation,
- use appropriate tools and techniques to gather, analyze, and interpret data,
- develop descriptions, explanations, predictions, and models using evidence,
- think critically and logically to make relationships between evidence and explanations,
- recognize and analyze alternative explanations and predictions, and
- communicate scientific procedures and explanations.

This activity also provides all students opportunities to develop understanding about inquiry and the nature of science as described in the *National Science Education Standards*. Specifically, it introduces the following concepts:

- Different kinds of questions suggest different kinds of scientific investigations.
- Current scientific knowledge and understanding guide scientific investigations.
- Technology used to gather data enhances accuracy and allows scientists to analyze and quantify results of investigations.
- Scientific explanations emphasize evidence, have logically consistent arguments, and use scientific principles, models, and theories.
- Science distinguishes itself from other ways of knowing and from other bodies of knowledge through the use of empirical standards, logical arguments, and skepticism, as scientists strive for the best possible explanations about the natural world.

Science Background for Teachers

The pursuit of scientific explanations often begins with a question about a natural phenomenon. Science is a way of developing answers, or improving explanations, for observations or events in the natural world. The scientific question can emerge from a child's curiosity about where the dinosaurs went or why the sky is blue. Or the question can extend scientists' inquiries into the process of extinction or the chemistry of ozone depletion.

Once the question is asked, a process of scientific inquiry begins, and there eventually may be an answer or a proposed explanation. Critical aspects of science include curiosity and the freedom to pursue that curiosity. Other attitudes and habits of mind that characterize scientific inquiry and the activities of scientists include intelligence, honesty, skepticism, tolerance for ambiguity, openness to new knowledge, and the willingness to share knowledge publicly.

Scientific inquiry includes systematic approaches to observing, collecting information, identifying significant variables, formulating and testing hypotheses, and taking precise, accurate, and reliable measurements. Understanding and designing experiments are also part of the inquiry process.

Scientific explanations are more than the results of collecting and organizing data. Scientists also engage in important processes such as constructing laws, elaborating models, and developing hypotheses based on data. These processes extend, clarify, and unite the observations and data and, very importantly, develop deeper and broader explanations. Examples include the taxonomy of organisms, the periodic table of the elements, and theories of common descent and natural selection.

One characteristic of science is that many explanations continually change. Two types of changes occur in scientific explanations: new explanations are developed, and old explanations are modified.

Just because someone asks a question about an object, organism, or event in nature does not necessarily mean that person is pursuing a scientific explanation. Among the conditions that must be met to make explanations scientific are the following:

• *Scientific explanations are based on empirical observations or experiments.* The appeal to authority as a valid explanation does not meet the requirements of science. Observations are based on sense experiences or on an extension of the senses through technology.

• *Scientific explanations are made public.* Scientists make presentations at scientific meetings or publish in professional journals, making knowledge public and available to other scientists.

• *Scientific explanations are tentative.* Explanations can and do change. There are no scientific truths in an absolute sense.

• *Scientific explanations are historical.* Past explanations are the basis for contemporary explanations, and those, in turn, are the basis for future explanations.

• *Scientific explanations are probabilistic.* The statistical view of nature is evident implicitly or explicitly when stating scientific predictions of phenomena or explaining the likelihood of events in actual situations.

• *Scientific explanations assume cause-effect relationships.* Much of science is directed toward determining causal relationships and developing explanations for interactions and linkages between objects, organisms, and events. Distinctions among causality, correlation, coincidence, and contingency separate science from pseudoscience.

• *Scientific explanations are limited.* Scientific explanations sometimes are limited by technology, for example, the resolving power of microscopes and telescopes. New technologies can result in new fields of inquiry or extend current areas of study. The interactions between technology and advances in molecular biology and the role of technology in planetary explorations serve as examples.

Science cannot answer all questions. Some questions are simply beyond the parameters of science. Many questions involving the meaning of life, ethics, and theology are examples of questions that science cannot answer. Refer to the *National Science Education Standards* for Science as Inquiry (pages 145-148 for grades 5-8 and pages 175-176 for grades 9-12), History and Nature of Science Standards (pages 170-171 for grades 5-8 and pages 200-204 for grades 9-12), and Unifying Concepts and Processes (pages 116-118). Chapter 3 of this document also contains a discussion of the nature of science.

Materials and Equipment

• 1 cube for each group of four students (black-line masters are provided).

(Note: you may wish to complete the first portion of the activity as a demonstration for the class. If so, construct one large cube using a cardboard box. The sides should have the same numbers and markings as the black-line master.)

• 10 small probes such as tongue depressors or pencils.

• 10 small pocket mirrors.

Instructional Strategy

Engage Begin by asking the class to tell you what they know about how scientists do their work. How would they describe a scientific investigation? Get students thinking about the process of scientific inquiry and the nature of science. This is also an opportunity for you to assess their current understanding of science. Accept student answers and record key ideas on the overhead or chalkboard.

Explore (The first cube activity can be done as a demonstration if you construct a large cube and place it in the center of the room.) First, have the students form groups of three or four. Place the cubes in the center of the table where the students are working. The students should not touch, turn, lift, or open the cube. Tell the students they have to identify a question associated with the cube. Allow the students to state their questions. Likely questions include:

- What is in the cube?
- What is on the bottom of the cube?
- What number is on the bottom?

You should direct students to the general question, *what is on the bottom of the cube*? Tell the students that they will have to answer the question by proposing an explanation, and that they will have to convince you and other students that their answer is *based on evidence*. (Evidence refers to observations the group can make about the visible sides of the cube.) Allow the students time to explore the cube and to develop answers to their question. Some observations or statements of fact that the students may make include:

- The cube has six sides.
- The cube has five exposed sides.
- The numbers and dots are black.
- The exposed sides have numbers 1, 3, 4, 5, and 6.
- The opposite sides add up to seven.
- The even-numbered sides are shaded.
- The odd-numbered sides are white.

Ask the students to use their observations (the data) to propose an answer to the question: *What is on the bottom of the cube*? The student groups should be able to make a statement such as: *We conclude there is a 2 on the bottom*. Students should present their reasoning for this conclusion. For example, they might base their conclusion on the observation that the exposed sides are 1, 3, 4, 5, and 6, and because 2 is missing from the sequence, they conclude it is on the bottom.

Use this opportunity to have the students develop the idea that combining two different but logically related observations creates a stronger explanation. For example, 2 is missing in the sequence (that is, 1, _, 3, 4, 5, 6) and that opposite sides add up to 7 (that is, 1—6; 3—4; _—5) and because 5 is on top, and 5 and 2 equal 7, 2 could be on the bottom.

If done as a demonstration, you might put the cube away without showing the bottom or allowing students to dismantle it. Explain that scientists often are uncertain about their proposed answers, and often have no way of knowing the absolute answer to a scientific question. Examples such as the exact ages of stars and the reasons for the extinction of prehistoric organisms will support the point.

Explain Begin the class period with an explanation of how the activity simulates scientific inquiry and provides a model for science. Structure the discussion so students make the connections between their experiences with the cube and the key points (understandings) you wish to develop.

Key points from the *Standards* include the following:

- Science originates in questions about the world.
- Science uses observations to construct explanations (answers to the questions). The more observations you had that supported your proposed explanation, the stronger your explanation, even if you could not confirm the answer by examining the bottom of the cube.
- Scientists make their explanations public through presentations at professional meetings and journals.
- Scientists present their explanations and critique the explanations proposed by other scientists.

The activity does not explicitly describe "the scientific method." The students had to work to answer the question and probably did it in a less than systematic way. Identifiable elements of a method—such as observation, data, and hypotheses—were clear but not applied systematically. You can use the experiences to point out and clarify scientific uses of terms such as observation, hypotheses, and data.

For the remainder of the second class period you should introduce the "story" of an actual scientific discovery. Historic examples such as Charles Darwin would be ideal. You could also assign students to prepare brief reports that they present.

Elaborate The main purpose of the second cube is to extend the concepts and skills introduced in the earlier activities and to introduce the role of prediction, experiment, and the use of technology in scientific inquiry. The problem is the same as the first cube: *What is on the bottom of the cube?* Divide the class into groups of three and instruct them to make observations and propose an answer about the bottom of the cube. Student groups should record their factual statements about the second cube. Let students identify and organize their observations. If the students are becoming too frustrated, provide helpful suggestions. Essential data from the cube include the following (see black-line master):

- Names and numbers are in black.
- Exposed sides have either a male or female name.
- Opposing sides have a male name on one side and a female name on the other.
- Names on opposite sides begin with the same letters.
- The number in the upper-right corner of each side corresponds to the number of letters in the name on that side.
- The number in the lower-left corner of each side corresponds to the number of the first letter that the names on opposite sides have in common.
- The number of letters in the names on the five exposed sides progresses from three (Rob) to seven (Roberta).

Four names, all female, could be on the bottom of the cube: Fran, Frances, Francene, and Francine. Because there are no data to show the exact name, groups might have different hypotheses. Tell the student groups that scientists use patterns in data to make predictions and then design an experiment to assess the accuracy of their prediction. This process also produces new data.

Tell groups to use their observations (the data) to make a prediction of the number in the upper-right corner of the bottom. The predictions will most likely be 4, 7, or 8. Have the team decide which corner of the bottom they wish to inspect and why they wish to inspect it. The students might find it difficult to determine which corner they should inspect. Let them struggle with this and even make a mistake—this is part of science! Have one student obtain a utensil, such as a tweezers, probe, or tongue depressor, and a mirror. The student may lift the designated corner less than one inch and use the mirror to look under the corner. This simulates the use of technology in a scientific investigation. The groups should describe the data they gained by the "experiment." Note that the students used technology to expand their observations and understanding about the cube, even if they did not identify the corner that revealed the most productive evidence.

If students observe the corner with the most productive information, they will discover an 8 on the bottom. This observation will confirm or refute the students' working hypotheses. Francine or Francene are the two possible names on the bottom. The students propose their answer to the question and design another experiment to answer the question. Put the cube away without revealing the bottom. Have each of the student groups present brief reports on their investigation.

Evaluate The final cube is an evaluation. There are two parts to the evaluation. First, in groups of three, students must create a cube that will be used as the evaluation exercise for other groups. After a class period to develop a cube, the student groups should exchange cubes. The groups should address the same question: *What is on the bottom of the cube?* They should follow the same rules—for example, they cannot pick up the cube. The groups should prepare a written report on the cube developed by their peers. (You may have the students present oral reports using the same format.) The report should include the following:

- title,
- the question they pursued,
- observation—data,
- experiment—new data,

- proposed answer and supporting data,
- a diagram of the bottom of the cube, and
- suggested additional experiments.

Due to the multiple sources of data (information), this cube may be difficult for students. It may take more than one class period, and you may have to provide resources or help with some information.

Remember that this activity is an evaluation. You may give some helpful hints, especially for information, but since the evaluation is for inquiry and the nature of science you should limit the information you provide on those topics.

Student groups should complete and hand in their reports. If student groups cannot agree, you may wish to make provisions for individual or "minority reports." You may wish to have groups present oral reports (a scientific conference). You have two opportunities to evaluate students on this activity: you can evaluate their understanding of inquiry and the nature of science as they design a cube, and you can assess their abilities and understandings as they figure out the unknown cube.

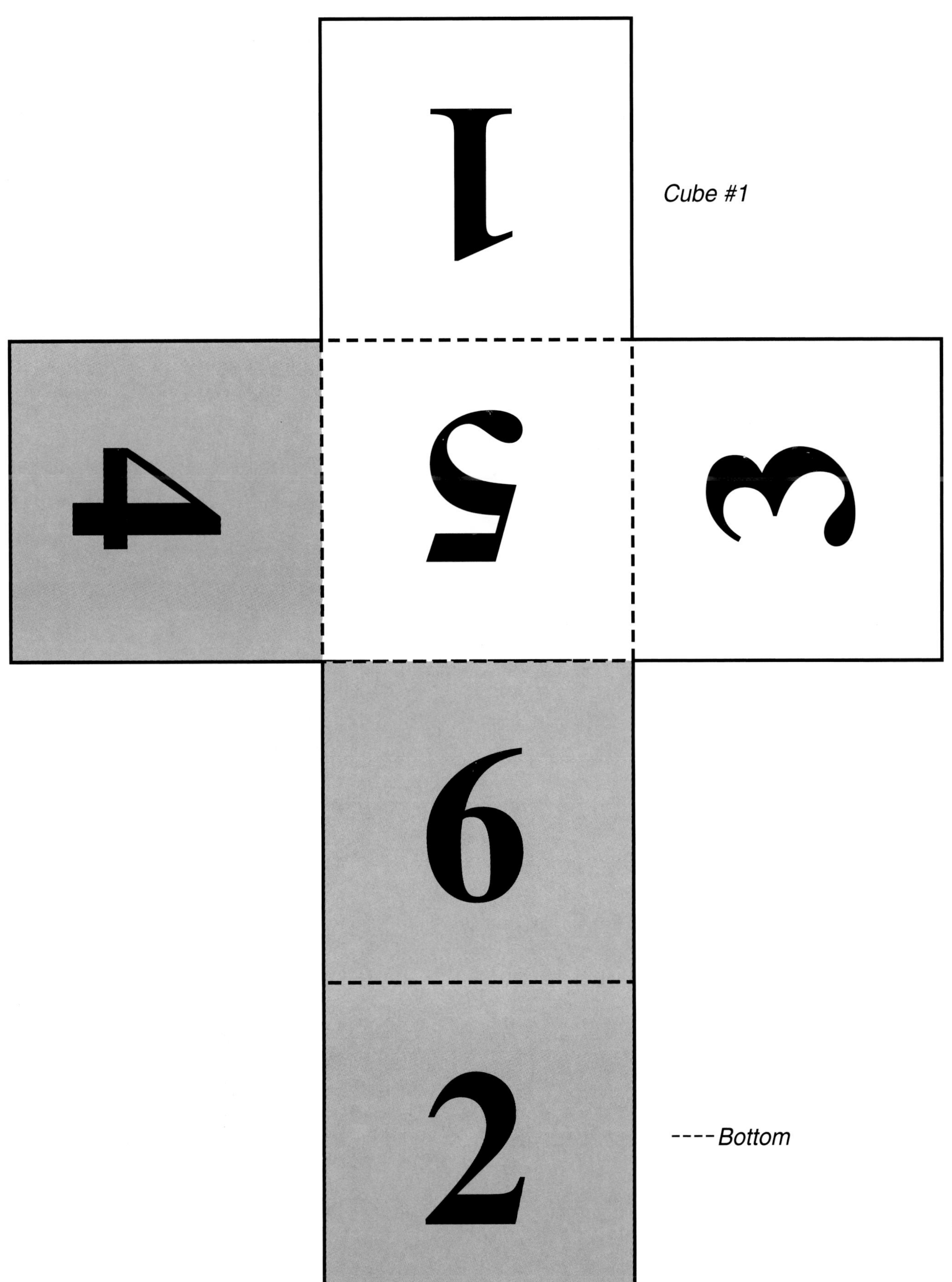
1
Cube #1
4
5
3
6
2
----Bottom

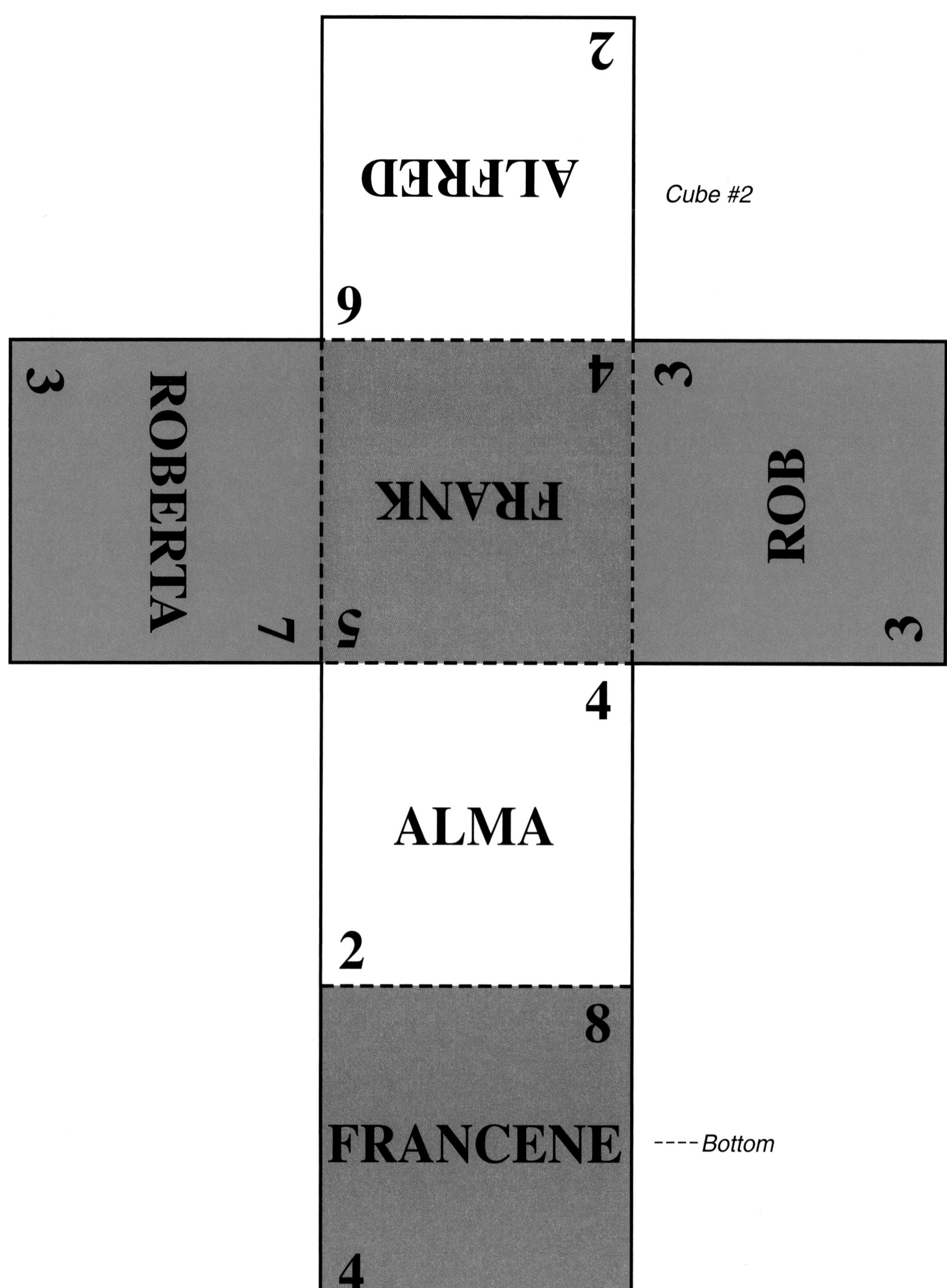
2
ALFRED
6
Cube #2
3
ROBERTA
7
4
FRANK
5
3
ROB
3
4
ALMA
2
8
FRANCENE
4
----Bottom

Cube #3

ACTIVITY 2

The Formulation of Explanations: An Invitation to Inquiry on Natural Selection

This activity uses the concept of natural selection to introduce the idea of formulating and testing scientific hypotheses. Through a focused discussion approach, the teacher provides information and allows students time to think, interact with peers, and propose explanations for observations described by the teacher. The teacher then provides more information, and the students continue their discussion based on the new information. This activity will help students in grades 5 through 8 develop several abilities related to scientific inquiry and formulate understandings about the nature of science as presented in the *National Science Education Standards*. This activity is adapted with permission from *BSCS: Biology Teachers' Handbook.*[3]

Standards-Based Outcomes

This activity provides all students with opportunities to develop the abilities of scientific inquiry as described in the *National Science Education Standards*. Specifically, it enables them to:

- identify questions that can be answered through scientific investigations,
- design and conduct a scientific investigation,
- use appropriate tools and techniques to gather, analyze, and interpret data,
- develop descriptions, explanations, predictions, and models using evidence,
- think critically and logically to make relationships between evidence and explanations,
- recognize and analyze alternative explanations and predictions, and
- communicate scientific procedures and explanations.

This activity also provides all students opportunities to develop understandings about inquiry, the nature of science, and biological evolution as described in the *National Science Education Standards*. Specifically, it conveys the following concepts:

- Different kinds of questions suggest different kinds of scientific investigations.
- Current scientific knowledge and understanding guide scientific investigations.
- Technology used to gather data enhances accuracy and allows scientists to analyze and quantify results of investigations.
- Scientific explanations emphasize evidence, have logically consistent arguments, and use scientific principles, models, and theories.
- Species evolve over time. Evolution is the consequence of the interactions of (1) the potential for a species to increase its numbers, (2) the genetic variability of offspring due to mutation and recombination of genes, (3) a finite supply of the resources required for life, and (4) the ensuing selection of those offspring better able to survive and leave offspring in a particular environment.

Science Background for Teachers

Many biological theories can be thought of as developing in five interrelated and overlapping stages. The first is a period of extensive observation of nature or analyzing the results of experiments. Darwin's observations would be an example of the former. Second, these observations lead scientists to ponder questions of "how" and "why." In the course of answering these questions, scientists infer explanations or make conjectures as working hypotheses. Third, in most cases, scientists submit hypotheses to formal, rigorous tests to check the validity of the hypotheses. At this point the hypotheses can be confirmed, falsified and rejected (not supported with evidence), or modified based on the evidence. This is a stage of experimentation. Fourth, scientists propose formal explanations by making public presentations at professional meetings or publishing their results in peer-reviewed journals. Finally, adoption of an explanation is recognized by other scientists as they begin referring to and using the explanation in their research and publications.

This activity focuses on the second and third stages in this brief summary of the development of biological theories. Chapters 2 and 3 of this document provide further discussion of these points. Review the "History and Nature of Science" and "Science as Inquiry" sections of the *National Science Education Standards* for further background on scientific investigations.

Materials and Equipment

None required.

Instructional Strategy

Engage Have the students work in groups of two or three. Begin by engaging the students with the problem and the basic information they will need to formulate a hypothesis.

TO THE STUDENTS: A farmer was working with dairy cattle at an agricultural experiment station. The population of flies in the barn where the cattle lived was so large that the animals' health was affected. So the farmer sprayed the barn and the cattle with a solution of insecticide A. The insecticide killed nearly all the flies.

Sometime later, however, the number of flies was again large. The farmer again sprayed with the insecticide. The result was similar to that of the first spraying. Most, but not all, of the flies were killed.

Again within a short time the population of flies increased, and they were again sprayed with the insecticide. This sequence of events was repeated five times; then it became apparent that insecticide A was becoming less and less effective in killing the flies.

Explore Imagine that the farmer consulted a group of student researchers. Have the student groups discuss the problem and prepare several different hypotheses to account for the observations. They should share their results with the class. Students might propose explanations similar to the following:

1. Decomposition of insecticide A with age.
2. The insecticide is effective only under certain environmental conditions—for example, certain temperatures and levels of humidity—which changed in the course of the work.
3. The flies genetically most susceptible to the insecticide were selectively killed. (This item should not be elicited at this point or developed if suggested.)

TO THE STUDENTS: One farmer noted that one large batch of the insecticide solution had been made and used in all the sprayings. Therefore, he suggested the possibility that the insecticide solution decomposed with age.

Have the student groups suggest at least two different approaches to test this hypothesis. The students may propose that investigation of several different predictions of a hypothesis contributes to the reliability of the conclusions drawn. In the present instance, one approach would be to use sprays of different ages on different populations of flies. A quite different approach would consist simply of making a chemical analysis of fresh and old solutions to determine if changes had occurred.

TO THE STUDENTS: The student researchers made a fresh batch of insecticide A. They used it instead of the old batch on the renewed fly population at the farmer's barn. Nevertheless, despite the freshness of the solution, only a few of the flies died.

The same batch of the insecticide was then tried on a fly population at another barn several miles away. In this case, the results were like those originally seen at the experiment station—that is, most of the flies were killed. Here were two quite different results with a fresh batch of insecticide. Moreover, the weather conditions at the time of the effective spraying of the distant barn were the same as when the spray was used without success at the experiment station.

Stop and have the student groups analyze the observations and list the major components of the problem and subsequent hypotheses. They might list what they know, what they propose as explanations, and what they could do to test their explanations. Students might identify the following:

1. Something about the insecticide.
2. The conditions under which the insecticide was used.
3. The way in which the insecticide was used.

4. The organisms on which the insecticide was used.

TO THE STUDENTS: So far our hypotheses have had to do with just a few of these components. Which ones?

The hypotheses so far have concerned only "something about the insecticide" and "the condition under which the insecticide was used," items 1 and 2 above.

TO THE STUDENTS: The advantage of analyzing a problem, as we have done in our list, consists in the fact that it lets us see what possibilities we have not considered.

What possibilities in the list have we not considered in forming our hypotheses?

Item 3, "the way in which the insecticide was used," may be pursued as a further exercise if the teacher so wishes. However, emphasis should be placed on Item 4, "the organisms on which the insecticide was used." This item is developed next.

Explain TO THE STUDENTS: Let us see if we can investigate the interactions between insecticide A and the flies. From your knowledge of biology, think of something that might have happened within the fly population that would account for the decreasing effectiveness of insecticide A.

The students may need help here, even if they have learned something about evolution and natural selection. Here is one way to help:

Ask the students to remember that after the first spraying, most, *but not all*, of the flies were killed. Ask them where the new population of flies came from—that is, who were the parents of the next generation of flies? Were the parents among the flies more susceptible or more resistant to the effects of insecticide A? Then remind them that the barn was sprayed again. If there are differences in the population to insecticide A susceptibility, which individuals would be more likely to survive this spraying? Remind them that dead flies do not produce offspring—only living ones can. The students might thus be led to see that natural selection, in this case in an imposed environment (the presence of the insecticide), might have resulted in the survival of only those individuals that were best adapted to live in the new environment (one with the insecticide). Because this activity centers on the formulation of explanations, it is important to introduce students to the scientific process they have been using. Following is a discussion from the *National Science Education Standards* that can serve as the basis for the explanation phase of the activity.

Evidence, Models, and Explanation[4]

Evidence consists of observations and data on which to base scientific explanations. Using evidence to understand interactions allows individuals to predict changes in natural and designed systems.

Models are tentative schemes or structures that correspond to real objects, events, or classes of events, and that have explanatory power. Models help scientists and engineers understand how things work. Models take many forms, including physical objects, plans, mental constructs, mathematical equations, and computer simulations.

Scientific explanations incorporate existing scientific knowledge and new evidence from observations, experiments, or models into internally consistent, logical statements. Different terms, such as "hypothesis," "model," "law," "principle," "theory," and "paradigm," are used to describe various types of scientific explanations. As students develop and as they understand more science concepts and processes, their explanations should become more sophisticated. That is, their scientific explanations should more frequently include a rich scientific knowledge base, evidence of logic, higher levels of analysis, greater tolerance of criticism and uncertainty, and a clearer demonstration of the relationship between logic, evidence, and current knowledge.

Elaborate Give the students a new problem—for example one of the investigations from *The Beak of the Finch*[5] or *Darwin's Dreampond.*[6] Have them work in groups to propose an explanation. The students should emphasize the role of hypotheses in the development of scientific explanations.

Evaluate Have the students consider the following case. Suppose a group of farmers notices the gradual acquisition of resistance to insecticide A over a period of months. They locate two other equally powerful although chemically unrelated insecticides, insecticides B and C. The local Agriculture Department sets up a program whereby all the farmers in the state will use only insecticide A for the current year. No one is to use insecticides B or C. The following year, everyone is directed to use insecticide B rather than insecticide A. The fly population, which had become resistant to insecticide A, is now susceptible to insecticide B and can be kept under control much more thoroughly than if the farmers had continued using insecticide A. At the beginning of the third year, all of the farmers begin using insecticide C, which again reduces the fly population greatly. As the fourth year begins, insecticide A is again used, and it proves to once again be extremely effective against the flies.

Have students analyze this situation and propose an explanation of what has happened. How would they design an investigation to support or reject their hypothesis?

ACTIVITY 3

Investigating Natural Selection

In this activity, the students experience one mechanism for evolution through a simulation that models the principles of natural selection and helps answer the question: How might biological change have occurred and been reinforced over time? The activity is designed for grades 9 through 12 and requires three class periods. This activity is adapted with permission from *BSCS Biology: A Human Approach.*[7]

Standards-Based Outcomes

This activity provides all students opportunities to develop understandings of biological evolution as described in the *National Science Education Standards*. Specifically, it conveys the concepts that:

• Species evolve over time. Evolution is the consequence of the interaction of (1) the potential for a species to increase in number, (2) the genetic variability of offspring due to mutation and recombination of genes, (3) a finite supply of the resources required for life, and (4) the ensuing selection of those offspring better able to survive and leave offspring in a particular environment. Item 4 is the primary emphasis of this activity. Teachers can introduce the other factors as appropriate.

• Natural selection and its evolutionary consequences provide a scientific explanation for the fossil record of ancient life forms, as well as for the striking molecular similarities observed among the diverse species of living organisms.

• Some living organisms have the capacity to produce populations of almost infinite size, but environments and resources are finite. The fundamental tension has profound effects on the interactions among organisms.

Science Background for Teachers

Many students have difficulty with the fundamental concepts of evolution. For example, some students express misconceptions about natural selection because they do not understand the relationship between variations within a population and the potential effect of those variations as the population continues to grow in numbers in an environment with limited resources. This is a dynamic understanding that derives from the four ideas presented in the learning outcomes for this activity.

This activity emphasizes natural selection. In particular, it presents students with the predator-prey relationship as one example of how natural selection operates in nature.

Students should understand that the process of evolution has two steps, referred to as genetic variation and natural selection. The first step is the development of genetic variation through changes such as genetic recombination, gene flow, and mutations. The second step, and the point of this activity, is selection. Differential survival and reproduction of organisms is due to a variety of environmental factors such as predator-prey relationships, resource shortages, and change of habitat. In any generation only a small percentage of organisms survives. Survival depends on an organism's genetic constitution that will, given circumstances such as limited resources, give a greater probability of survival and reproduction.[8]

Materials and Preparation (per class of 32)

8 petri dish halves

8 36- x 44-in. pieces of fabric, 4 each of 2 different patterns

8 sheets of graph paper

8 zip-type plastic sandwich bags containing 120 paper dots, 20 each of 6 colors (labeled "Beginning Population")

8 sets of colored pencils with colors similar to the paper dot colors

8 zip-type plastic sandwich bags of spare paper dots in all colors

watch or clock with a second hand

computer with spreadsheet software program (optional)

24 forceps (optional)

Choose fabric patterns that simulate natural environments, such as floral, leaf, or fruit prints. The patterns should have several colors and be of intricate design; small prints work better than large blocky prints. Select two designs, each with a different predominant color. Label one design Fabric A and the other Fabric B. The use of two designs enables the students to demonstrate the evolution of different color types from the same starting population.

Use a paper punch to punch out quarter-inch paper dots from construction paper of six different colors. Select two light colors (including white) and two dark colors so that they will compete against each other. Include at least two colors that blend well with the fabrics. For each color, put 100 dots into each of 8 zip-type plastic sandwich bags. Put 20 dots of each color (for a total of 120 dots of 6 colors) into each of 24 additional bags. Label these bags "Beginning Population." Enlist student aides or ask for student volunteers to punch dots or stuff bags at home or after school. As an alternative to paper dots, you might try colored aquarium gravel or colored rice. Both are heavier than paper dots and are less likely to blow around the room. You could color the rice grains with food dyes according to the criteria specified above for the dots. You also might use gift-wrapping paper instead of the pieces of fabric.

Instructional Strategy

Engage Begin by asking students what they know about the theory of natural selection. Ask them what predator-prey relationships have to do with biological evolution. Use the discussion as a means to have them explain how they think evolution occurs and the role of predator-prey relations in the process. At this point in the lesson, accept the variety of student responses, and determine any misconceptions the students express. You could present a historical example (see the discussion of fossils in chapter 3 of this volume) or an example from *The Beak of the Finch* by Jonathan Weiner or *Darwin's Dreampond* by Tijs Goldschmidt.

Because the instructional procedures are complex for this activity, you will have to be fairly explicit about the process. Tell the students they will work in teams of four. (If your class does not divide evenly, use teams of five). The activity calls for half of the teams to use Fabric A and half of the teams to use Fabric B. It will help if you go through a "trial run" before students begin the activity.

Explore Step 1. Tell the students to pick a "game warden" from each group of four. The other group members will be the predators.

Step 2. Examine the paper dots in the bags labeled "Starting Population" and record the number of individuals (dots) of each color. All of the dots represent individuals of a particular species, and the individuals can be one of six colors.

Step 3. Make certain that half of the teams use Fabric A and half use Fabric B. The procedures remain the same for both groups.

Steps 4 and 5. Tell the predators to turn away from the habitats. The game warden then spreads one of the bags of "Beginning Population" across the fabric and tells the predators to turn around and gather prey—i.e., the dots. The predators must stop hunting (picking up dots) when the game warden says "Stop" in 20 seconds. If the predators have difficulty picking up the paper dots, provide forceps.

Step 6. After the hunting is stopped, the students should carefully collect all of the dots that remain on the fabric and sort them by color. The game wardens are responsible for recording these data on the graph paper using the colored pencils corresponding to the dot colors.

Step 7. To simulate reproduction among the paper dots, add three paper dots for each remaining dot of that color. These paper dots, obtained from the bags containing extra dots, represent offspring.

Step 8. Repeat the predation using the second generation of dots. Again record the number of remaining dots in the second generation.

Step 9. Explain to the students that they do not have to simulate reproduction as they did before, but rather should calculate the number of individuals that would be in the third-generation beginning population.

Step 10. The construction and analysis of bar graphs is a critical and time-consuming part of this activity. Place the color of survivors on the horizontal axis and the number of the beginning population (or second generation) on the vertical axis of this activity. If you have ready access to computers and spreadsheet programs, you could incorporate the use of spreadsheets during this step.

Explain Step 11. Study the bar graphs of each generation. Discuss the following questions (possible student responses are included).

• Which, if any, colors of paper dots survived better than others in the second- and third-generation beginning populations of paper dots?

Answers will vary depending on the color of the fabric that the students used. The beginning populations for the second and third generations should include more dots that are of colors similar to the fabric and fewer dots that are of colors that stand out against the fabric. The change between the first and third generations should be more dramatic than the change between the first and second generations.

• What might be the reason that predators did not select these colors as much as they did other colors?

Some colors were better camouflaged than others—they blended into the environment.

• What effect did capturing a particular color dot have on the numbers of that color in the following generations?

When an individual is removed from a population and dies, in this case through predation, that individual no longer reproduces. The students should realize that heavy predation leads to a decrease in the size of the population and in the size of the gene pool.

Step 12. Allow the students enough time to re-sort the colored dots into the appropriate bags. Be sure the students recount the dots in each bag and replace missing dots. Have a three-hole punch and construction paper on hand to replace lost dots.

Elaborate This portion of the activity provides you with an opportunity to assess the learners' understanding of evolution and the mechanisms by which it occurs. Before the students begin to work on these tasks, display a piece of Fabric A and a piece of Fabric B and ask the learners to post their third generation bar graphs beside the fabric that they used. The learners now will benefit by comparing their results with those from other teams that used the same fabric as well as with those from teams that used a different fabric. These comparisons will give them more data with which to construct explanations for the results that they see.

1. How well do the class data support your team's conclusions in Step 11?

Students need to be able to analyze the relationship between their response in Step 11 and the cumulative data. The specific response should address the relationship between the team data and the class data.

2. Imagine a real-life predator-prey relationship and write a paragraph that describes how one or more characteristics of the predator population or the prey population might change as a result of natural selection.

The students should explain that variation exists in populations. Individuals with certain characteristics are better adapted than other individuals to their environment, and consequently survive to produce offspring; less well-adapted individuals do not. The offspring, in turn, possess characteristics similar to those of their parents, and that makes them better adapted to the environment as well. These two concepts are the basis of natural selection, and they explain how populations evolve.

Little variation in a population of organisms would mean that fewer differences would be expressed in the offspring. Fewer differences would mean that individuals would have similar advantages and disadvantages in the prevailing environmental conditions. This similarity, in turn, would mean that their survival and reproductive rates would be similar, so few heritable differences then would be passed on to the next generation.

Evaluate Have the students write one paragraph that summarizes their understanding of biological evolution. Refer to the learning outcomes and the *National Science Education Standards.* Expect that students will describe that in a population of organisms, variation exists among characteristics that parents pass on to their offspring. Individuals with certain characteristics might have a slight advantage over other individuals and thus live longer and reproduce more. If this advantage remains, the difference would be more noticeable over time. These changes could eventually lead to new species. The process of natural selection, then, provides an explanation for the relatedness of organisms and for biological change across time.

ACTIVITY 4

Investigating Common Descent: Formulating Explanations and Models

In this activity, students formulate explanations and models that simulate structural and biochemical data as they investigate the misconception that humans evolved from apes. The activities require two 45-minute periods. They are designed for use in grades 9 through 12. This activity is adapted with permission from *Evolution: Inquiries into Biology and Earth Science* by *BSCS*.[9]

Standards-Based Outcomes

This activity provides opportunities for all students to develop abilities of scientific inquiry as described in the *National Science Education Standards*. Specifically, it enables them to:

- formulate descriptions, explanations, predictions, and models using evidence,
- think critically and logically to make relationships between evidence and explanations, and
- recognize and analyze alternative explanations and predictions.

In addition, the activity provides all students opportunities to develop fundamental understandings in the life sciences as described in the *National Science Education Standards*. Specifically, it conveys the following concepts:

- In all organisms, the instructions for specifying the characteristics of the organism are carried in DNA, a large polymer formed from subunits of four kinds (A, G, C, and T). The chemical and structural properties of DNA explain how the genetic information that underlies heredity is both encoded in genes (as a string of molecular "letters") and replicated (by a templating mechanism).
- The millions of different species of plants, animals, and microorganisms that live on earth today are related by descent from common ancestors.
- Biological classifications are based on how organisms are related. Organisms are classified into a hierarchy of groups and subgroups based on similarities that reflect their evolutionary relationships. The species is the most fundamental unit of classification.

Science Background for Teachers

One of the most common misconceptions about evolution is seen in the statement that "humans came from apes." This statement assumes that organisms evolve through a step-by-step progression from "lower" forms to "higher" forms of life and the direct transformation of one living species into another. Evolution, however, is not a progressive ladder. Furthermore, modern species are derived from, but are not the same as, organisms that lived in the past.

This activity has extensive historical roots. Few question the idea that Charles Darwin's *Origin of Species* in 1859 produced a scientific revolution. In essence, Darwin proposed a constellation of ideas that included: organisms of different kinds descended from a common ancestor (common descent); species multiply over time (speciation); evolution occurs through gradual changes in a population (gradualism); and competition among species for limited resources leads to differential survival and reproduction (natural selection). This activity centers on the theory of common descent.

The theory of common descent was revolutionary because it introduced the concept of gradual evolution based on natural mechanisms. The theory of common descent also replaced a model of straight-line evolution with that of a branching model based on a single origin of life and subsequent series of changes—branching—into different species.

Based on his observations during the voyage of the H.M.S. *Beagle*, Darwin concluded that three species of mockingbirds on the Galapagos Islands must have some connection to the single species of mockingbird on the South American mainland. Here is the intellectual connection between observations and explanation. A species could produce multiple descendent species. Once this idea was realized, it was but a series of logical steps to the inferences that all birds, all vertebrates, and so on, had common ancestors.

Common descent has become a conceptual backbone for evolutionary biology. In large measure, this is so because common descent has significant explanatory power. Immediately, the idea found supporting evidence in comparative anatomy, comparative embryology, systematics, and biogeography. Recently, molecular biology has provided further support, as the students will discover in this activity. See Chapter 3 of this document and page 185 of the *National Science Education Standards* for more discussion of this topic.

This activity also introduces students to scientific evidence, models, and explanations as described in the accompanying excerpt drawn from the *National Science Education Standards.*

Evidence, Models, and Explanation[10]

Evidence consists of observations and data on which to base scientific explanations. Using evidence to understand interactions allows individuals to predict changes in natural and designed systems.

Models are tentative schemes or structures that correspond to real objects, events, or classes of events, and that have explanatory power. Models help scientists and engineers understand how things work. Models take many forms, including physical objects, plans, mental constructs, mathematical equations, and computer simulations.

Scientific explanations incorporate existing scientific knowledge and new evidence from observations, experiments, or models into internally consistent, logical statements. Different terms, such as "hypothesis," "model," "law," "principle," "theory," and "paradigm," are used to describe various types of scientific explanations. As students develop and as they understand more science concepts and processes, their explanations should become more sophisticated. That is, their scientific explanations should more frequently include a rich scientific knowledge base, evidence of logic, higher levels of analysis, greater tolerance of criticism and uncertainty, and a clearer demonstration of the relationship between logic, evidence, and current knowledge

Materials and Equipment

For each student:

- Notebook
- Pencil

For each group of four students

- 4 sets of black, white, green, and red paper clips, each set with 35 paper clips

For the entire class:

- Overhead transparencies of *Characteristics of Apes and Humans*, Table 1, and *Morphological Tree,* Figure 1
- Overhead projector

Instructional Strategy: Part I

Engage Ask the students: When you hear the word "evolution," what do you think of first? Have the students explain what they understand about evolution. For many people, the first thing that comes to mind is often the statement "Humans evolved from apes." Did humans evolve from modern apes, or do modern apes and humans have a common ancestor? Do you understand the difference between these two questions? This activity will give you the opportunity to observe differences and similarities in the characteristics of humans and apes. The apes discussed in this activity are the chimpanzee and the gorilla.

Explore Review Table 1, *Characteristics of Apes and Humans,* with the class. Make sure the students know that gibbons, chimpanzees, gorillas, and orangutans are four groups included in the ape family. Chimpanzees and gorillas represent the African side of the family; gibbons and orangutans represent the Asian side of the family. We focus only on the chimpanzee and gorilla in this activity. The only modern representative of the human family is *Homo sapiens*, although paleontologists have

Table 1.
Characteristics of Apes and Humans

Characteristics	Apes	Humans
Posture	Bent over or quadrupedal "knuckle-walking" common	Upright or bipedal
Leg and arm length	Arms longer than legs; arms adapted for swinging, usually among trees	Legs usually longer than arms; legs adapted for striding
Feet	Low arches; opposable big toes, capable of grasping	High arches; big toes in line with other toes; adapted for walking
Teeth	Prominent teeth; large gaps between canines and nearby teeth	Reduced teeth; gaps reduced or absent
Skull	Bent forward from spinal column; rugged surface; prominent brow ridges	Held upright on spinal column; smooth surface
Face	Sloping; jaws jut out; wide nasal opening	Vertical profile; distinct chin; narrow nasal opening
Brain size	280 to 705 cm^3 (living species)	400 to 2000 cm^3 (fossil to present)
Age at puberty	Usually 10 to 13 years	Usually 13 years or older
Breeding season	Estrus at various times	Continual

Figure 1.
Evolutionary relationships among organisms derived from comparisons of skeletons and other characteristics

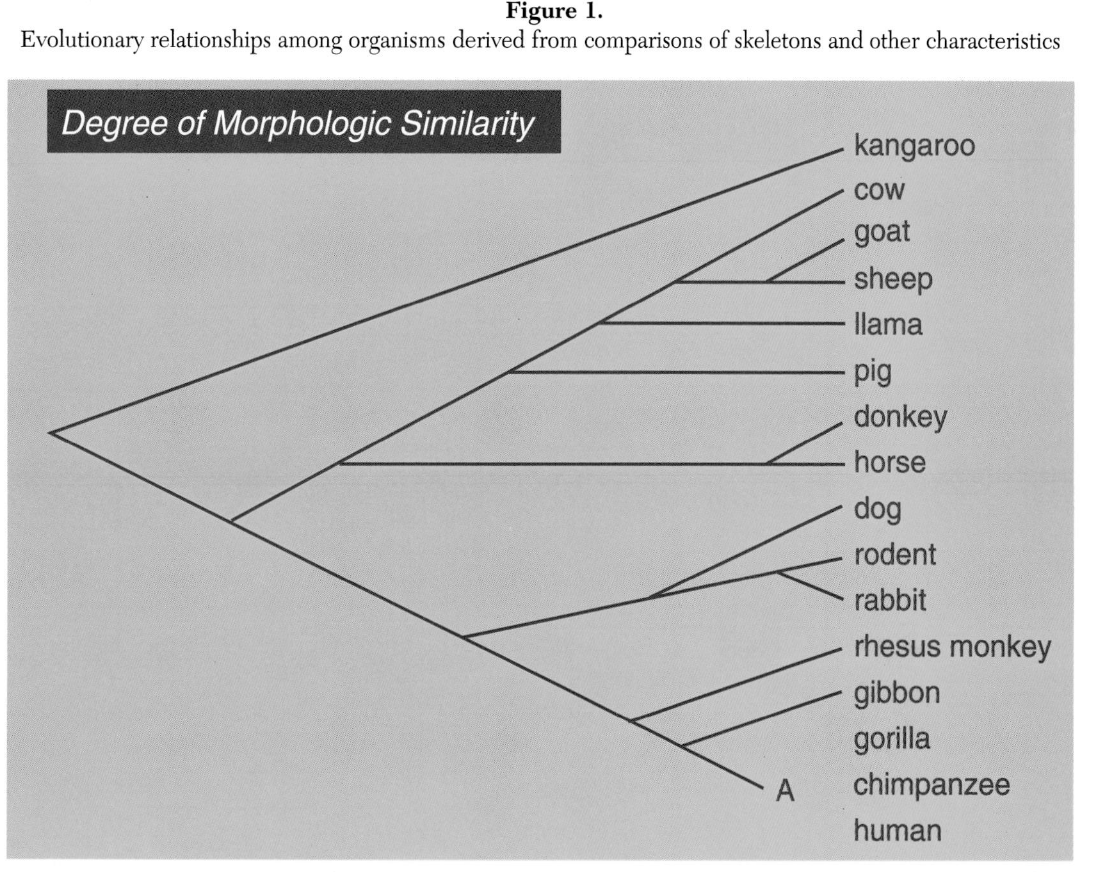

found fossil remains of other members, such as *Australopithecus afarensis* ("Lucy") and *Homo sapiens neandertalensis*.

Next discuss how the students can use the data to determine the relationships between humans, apes, and other animals. It might not be obvious that closely related organisms share more similarities than do distantly related organisms. Guide the students to the idea that structures might be similar because they carry out the same functions or because they were inherited from a common ancestor. Only those similarities that arise from a common ancestor can be used to determine evolutionary relationships.

Use the transparency of the *Morphological Tree*, Figure 1, for this discussion. Diagrams called branching trees illustrate relationships among organisms. One type of branching tree, called a morphological tree, is based on comparisons of skulls, jaws, skeletons, and other structures. Look carefully at the morphological tree.

Explain Ask the students to find the part of the morphological tree that shows the relationships between gorillas, chimpanzees, and humans. They will notice that there are no lines showing relationships. They should work with partners and develop three hypotheses to explain how these organisms are related. On a sheet of notebook paper, they should make a diagram of their hypotheses by drawing lines from Point A to each of the three organisms (G = gorilla, C = chimpanzee, H = human, A = common ancestor).

Allow the students to develop their own hypotheses. Give them help only if you see they are not making any progress. Three hypotheses the students might propose are shown below (although not necessarily in the same order).

Possible evolutionary relationships:

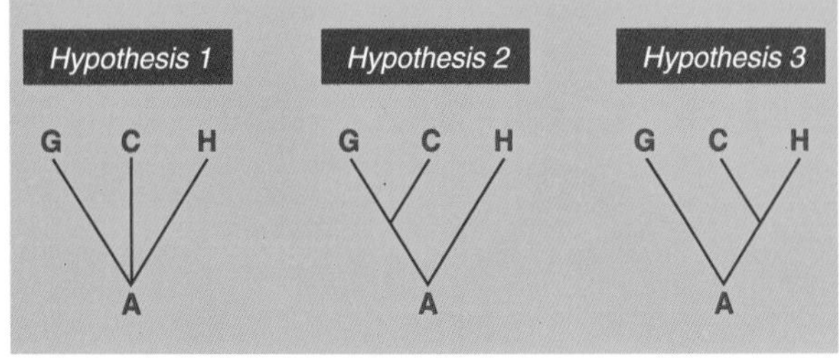

Instructional Strategy: Part II

Elaborate Modern research techniques allow biologists to compare the DNA that codes for certain proteins and to make predictions about the relatedness of the organisms from which they took the DNA. Students will use models of these techniques to test their hypotheses and determine which one is best supported by the data they develop.

Procedure Step 1. Working in groups of four, "synthesize" strands of DNA according to the following specifications. Each different color of paper clip represents one of the four bases of DNA:

black = adenine (A) green = guanine (G)
white = thymine (T) red = cytosine (C)

Students should synthesize DNA strands by connecting paper clips in the proper sequence according to specifications listed for each group member. When they have completed the synthesis, attach a label to Position 1 and lay your strands on the table with Position 1 on the left.

Each student will synthesize one strand of DNA. Thirty-five paper clips of each color should provide an ample assortment. To save time, make sure all strands are synthesized simultaneously. Emphasize to the students that they are using models to test the hypotheses they developed in the first part of the investigation. Following are directions for the respective groups:

- Group member 1

Synthesize a strand of DNA that has the following sequence:

Position 1 Position 20
A-G-G-C-A-T-A-A-A-C-C-A-A-C-C-G-A-T-T-A

Label this strand "human DNA." This strand represents a small section of the gene that codes for human hemoglobin protein.

- Group member 2

Synthesize a strand of DNA that has the following sequence:

Position 1 Position 20
A-G-G-C-C-C-C-T-T-C-C-A-A-C-C-G-A-T-T-A

Label this strand "chimpanzee DNA." This strand represents a small section of the gene that codes for chimpanzee hemoglobin protein.

• Group member 3
Synthesize a strand of DNA that has the following sequence:

Position 1 Position 20
A-G-G-C-C-C-C-T-T-C-C-A-A-C-C-A-G-G-C-C

Label this strand "gorilla DNA." This strand represents a small section of the gene that codes for gorilla hemoglobin protein.

• Group member 4
Synthesize a strand of DNA that has the following sequence:

Position 1 Position 20
A-G-G-C-C-G-G-C-T-C-C-A-A-C-C-A-G-G-C-C

Label this strand "common ancestor DNA." This DNA strand represents a small section of the gene that codes for the hemoglobin protein of a common ancestor of the gorilla, chimpanzee, and human.

(You will use this strand in Part III.) Emphasize to students that they will be using a model constructed from *hypothetical* data in the case of the common ancestor, since no such DNA yet exists, but that the other three sequences are real.

Step 2. Students should compare the human DNA to the chimpanzee DNA by matching the strands base by base (paper clip by paper clip).

Step 3. Students should count the number of bases that are not the same. Record the data in a table. Repeat these steps with the human DNA and the gorilla DNA.

The data for the hybridizations are as follows: chimpanzee DNA, 5 unmatched bases; gorilla DNA, 10 unmatched bases. Be sure to ask the students to save all of their DNA strands for Part III.

Evaluate 1. How do the gorilla DNA and the chimpanzee DNA compare with the human DNA?

The human DNA is more similar to the chimpanzee DNA than the gorilla DNA.

2. What do these data suggest about the relationship between humans, gorillas, and chimpanzees?

The data suggest that humans are more closely related to the chimpanzee than they are to the gorilla.

3. Do the data support any of your hypotheses? Why or why not?

Hybridization data for human DNA

Human DNA compared to:	Number of matches	Unmatched bases
Chimpanzee DNA		
Gorilla DNA		

Data for common ancestor DNA

Common ancestor DNA compared to:	Number of matches	Unmatched bases
Human DNA		
Chimpanzee DNA		
Gorilla DNA		

The data lend support to the hypothesis that the chimpanzee is more closely related to humans than the gorilla is.

4. What kinds of data might provide additional support for your hypotheses?

The students could test the hypotheses using additional data from DNA sequences or morphological features. They also could gather data from the fossil record.

Instructional Strategy: Part III

Begin this part by pointing out that biologists have determined that some mutations in DNA occur at a regular rate. They can use this rate as a "molecular clock" to predict when two organisms began to separate from a common ancestor. Most evolutionary biologists agree that humans, gorillas, and chimpanzees shared a common ancestor at one point in their evolutionary history. They disagree, however, on the specific relationships among these three species. In this part of the activity, you will use data from your paper-clip model to evaluate different hypotheses about the relationships between humans, gorillas, and chimpanzees.

Evolutionary biologists often disagree about the tempo of evolutionary change and about the exact nature of speciation and divergence. Reinforce the idea that models can be useful tools for testing hypotheses.

Procedure Step 1. Assume that the common ancestor DNA synthesized in Part II represents a section of the hemoglobin gene of a hypothetical common ancestor. Compare this common ancestor DNA to all three samples of DNA (gorilla, human, and chimpanzee), one sample at a time. Record the data in a table.

The data for the comparisons are as follows: human DNA, 10 unmatched bases; chimpanzee DNA, 8 unmatched bases; gorilla DNA, 3 unmatched bases.

Evaluate 1. Which DNA is most similar to the common-ancestor DNA?

Gorilla DNA is most similar to the common-ancestor DNA.

2. Which two DNAs were most similar in the way that they compared to the common-ancestor DNA?

Human DNA and chimpanzee DNA have similar patterns when compared to the common ancestor DNA.

3. Which of the hypotheses developed in Part I do your data best support?

Answers will vary.

4. Do your findings prove that this hypothesis is correct? Why or why not?

Data from the models do not *prove* the validity of a hypothesis, but they do provide some direction for additional research.

5. Based on the hypothesis that your data best supported, which of the following statements is most accurate? Explain your answer in a short paragraph.

(a) Humans and apes have a common ancestor.
(b) Humans evolved from apes.

The students should infer that humans and apes share a common ancestor, represented by a common branching point.

6. According to all the data collected, which of the following statements is most accurate? Explain your answer in a short paragraph.

(a) Chimpanzees and humans have a common ancestor.
(b) Chimpanzees are the direct ancestors of humans.

The students should infer that chimpanzees and humans share a common ancestor and that *modern* chimpanzees are not the direct *ancestors* of humans.

7. A comparison of many more DNA sequences indicates that human DNA and chimpanzee DNA are 98.8 percent identical. What parts of your data support this result?

The morphological tree and the DNA comparison data indicate that humans are closely related to chimpanzees.

8. What methods of science did you use in this activity?

Many answers are possible, including making observations, forming and testing hypotheses, and modeling.

ACTIVITY 5

Proposing Explanations for Fossil Footprints

In this activity, students observe and interpret "fossil footprint" evidence. From the evidence, they are asked to construct defensible hypotheses or explanations for events that took place in the geological past. The estimated time requirement for this activity is two class periods. This activity is designed for grades 5 through 8. The activity is adapted with permission from the Earth Science Curriculum Project.[11]

Standards-Based Outcomes

This activity provides all students an opportunity to develop the abilities of scientific inquiry and understanding of the nature of science as described in the *National Science Education Standards.* Specifically, it enables them to:

- propose explanations and make predictions based on evidence,
- recognize and analyze alternative explanations and predictions,
- understand that scientific explanations are subject to change as new evidence becomes available,
- understand that scientific explanations must meet certain criteria. First and foremost, they must be consistent with experimental and observational evidence about nature, and must make accurate predictions, when appropriate, about systems being studied. They should also be logical, respect the rules of evidence, be open to criticism, report methods and procedures, and make knowledge public. Explanations of how the natural world changes based on myths, personal beliefs, religious values, mystical inspiration, superstition, or authority may be personally useful and socially relevant, but they are not scientific.

Science Background for Teachers

This activity provides teachers with the opportunity to help students realize the differences between observations and inferences. In terms of the *Standards*, it centers on the development of explanations based on evidence. It encourages students to think critically about the inferences they make and about the logical relationships between cause and effect.

Observations or statements of observations should have agreement by all individuals: "These are fossil footprints," or "The dimensions of one of the footprints is 20 cm by 50 cm." Inferences are statements that propose possible explanations for observations: "The two sets of footprints represent a fight between the animals." If this is true, then what evidence could you look for to support the inference. Note that the primary emphasis for this activity is developing abilities and understandings for "Science as Inquiry" as described in the *Standards.*[12]

Materials and Equipment

- Make an overhead transparency of the footprint puzzle from the master provided on page 89. Have a blank piece of paper on hand to mask the puzzle when it is put on the projector.

Instructional Strategy

Engage Project position 1 of the footprints from the overhead by covering the other two positions with a blank piece of paper. Tell the students that tracks like these are common in parts of New England and in the southwestern United States. Point out to the students that they will be attempting to reconstruct happenings from the geological past by analyzing a set of fossilized tracks. Their problem is similar to that of a detective. They are to form defensible explanations of past events from limited evidence. As more evidence becomes available, their hypotheses must be modified or abandoned. The only clues are the footprints themselves. Ask the students: Can you tell anything about the size or nature of the organisms? Were all the tracks made at the same time? How many animals were involved? Can you reconstruct a series of events represented by this set of fossil tracks?

Have the students discuss each of the questions. Accept any reasonable explanations students offer. Try consistently to point out the difference between what they observe and what they infer. Ask them to suggest evidence that would support their proposed explanations.

Explore Reveal the second position of the puzzle and allow time for the students to consider the new information. Students will see that the first explanation may need to be modified and new ones added.

Next project the complete puzzle and ask students to interpret what happened. A key point for students to recognize is that any reasonable explanation must be based only on those proposed explanations that still apply when all of the puzzle is projected. Any interpretation that is consistent with all the evidence is acceptable.

Should it become necessary to challenge the students' thinking and stimulate the discussion, the following questions may help. Students should give evidence or suggest what they would look for as evidence to support their proposed explanations.

- In what directions did the animals move?
- Did they change their speed and direction?
- What might have changed the footprint pattern?
- Was the land level or irregular?
- Was the soil moist or dry on the day these tracks were made?
- In what kind of rock were the prints made?
- Were the sediments coarse or fine where the tracks were made?

The environment of the track area also should be discussed. If dinosaurs made the tracks, the climate probably was warm and humid. If students propose that some sort of obstruction prevented the animals from seeing each other, this might suggest vegetation. Or perhaps the widened pace might suggest a slope. Speculate on the condition of the surface at the time the footprints were made. What conditions were necessary for their preservation?

Explain An imaginative student should be able to propose several possible explanations. One of the most common is that two animals met and fought. No real reason exists to assume that one animal attacked and ate the other. Ask students who propose this explanation to indicate the evidence. If they could visit the site, what evidence would they look for that would support their explanation. Certain lines of evidence—the quickened gaits, circular pattern, and disappearance of one set of tracks—could support the fight explanation. They might, however, support an explanation of a mother picking up her baby. The description and temperament of the animals involved are open to question. Indeed, we lack the evidence to say that the tracks were made at the same time. The intermingling shown in the middle section of the puzzle may be evidence that both tracks were made at one time, but it could be only a coincidence. Perhaps one animal passed by and left, and then the other arrived.

Discuss the expected learning outcomes related to scientific inquiry and the nature of science. To answer the questions posed by the set of fossil footprints, the students, like scientists, constructed reasonable explanations based solely on their logical interpretation of the available evidence. They recognized and analyzed alternative explanations by weighing the evidence and examining the logic to decide which explanations seemed most reasonable. Although there may have been several plausible explanations, they did not all have equal weight. In a manner similar to the way scientists work, students should be able to use scientific criteria to find, communicate, and defend the preferred explanation.

Elaborate You can have more discussions on interpreting series of events using animal prints students find outdoors and reproduce for the class. Do not forget to look for human footprints. Have students design a different fossil footprint puzzle. Choose several different ones and have student teams repeat the activity using the same learning goals.

Evaluate Describe a specific event involving two or more people or animals where footprint evidence remains. Ask the students, either in teams or individually, to diagram footprint evidence that could lead to several different, yet defensible, explanations regarding what took place. They should be able to explain the strengths and weaknesses of each explanation using their footprint puzzle.

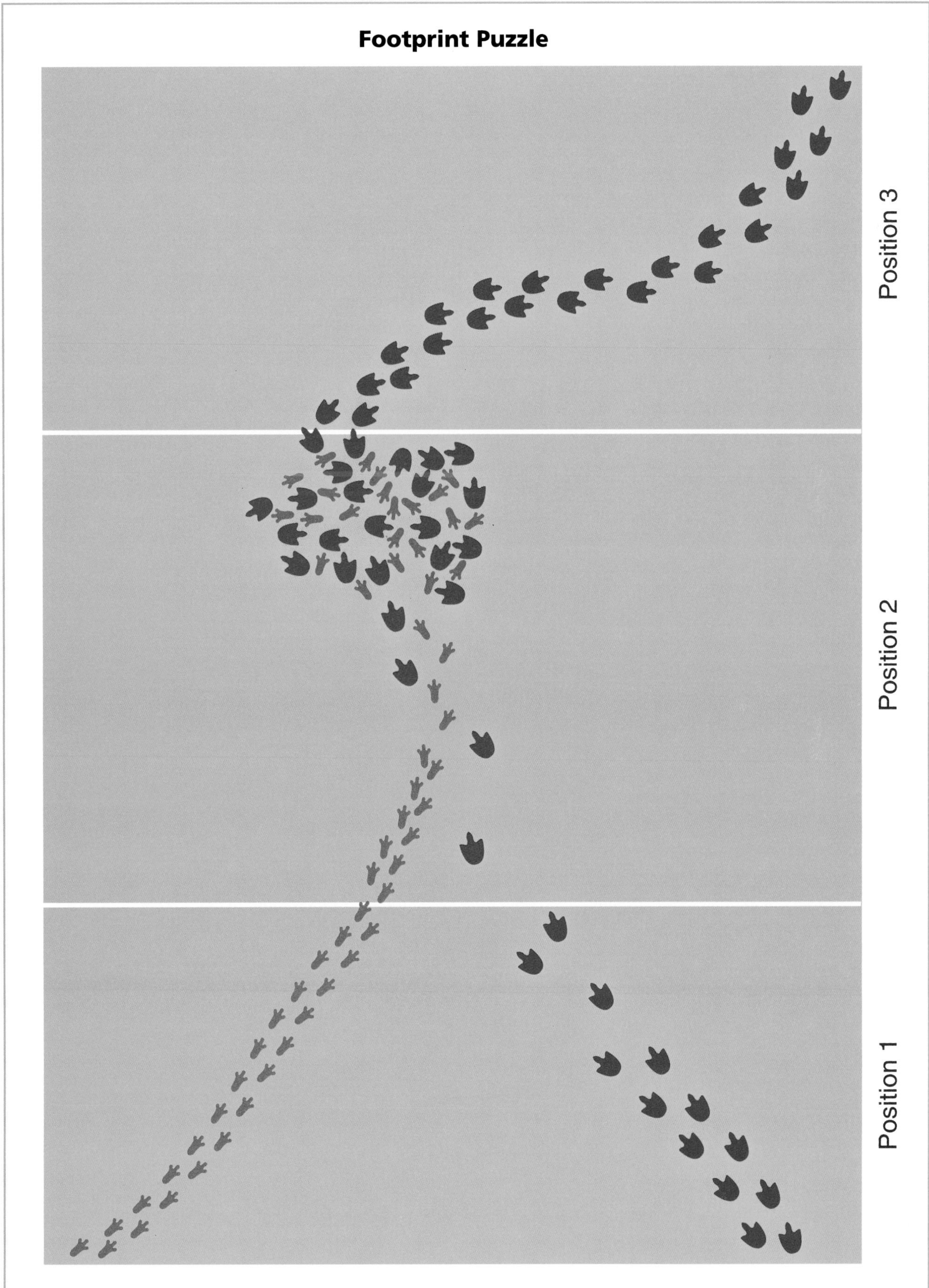
Footprint Puzzle
Position 3
Position 2
Position 1

ACTIVITY 6

Understanding Earth's Changes Over Time

Comparing the magnitude of geologic time with spans of time in a person's lifetime is difficult for many students. In this activity, students use a long paper strip and a reasonable scale to represent visually all of geologic time, including significant events in the development of life on earth as well as recent human events. The investigation requires two class periods and is appropriate for grades 5 through 12. The activity is adapted with permission from the Earth Science Curriculum Project.[13]

Standards-Based Outcomes

This activity provides all students with an opportunity to develop understandings of the earth systems as described in the *National Science Education Standards.* Specifically, it introduces them to the following concepts:

- A mathematical scale representing the length of geologic time.
- The relationship of time between human events, events in earth's history, and the total age of the earth.
- The formation of the sun, the earth, and the rest of the solar system from a nebular cloud of dust and gas 4.6 billion years ago.
- The estimation of geologic time by observing rock sequences and using fossils to correlate the sequences at various locations. A current method of dating earth materials uses the known decay rates of radioactive isotopes present in rocks to measure the time since the rock was formed.
- The ongoing evolution of the earth system resulting from interactions among the solid earth, the oceans, the atmosphere, and organisms.
- The evidence for one-celled forms of life—the bacteria—extending back more than 3.5 billion years. The evolution of life caused dramatic changes in the composition of the earth's atmosphere, which did not originally contain oxygen.

Science Background for Teachers

Geologic time is largely subdivided on the basis of the evolution of life and on the amount and type of crustal activity that occurred in the past. Geologic time is ordered both relatively and absolutely. For relative dating the sequence in which rock strata formed is important; to explain the complete time scale for all of geologic history required correlating rock formations throughout the world. Fossils are important guides in this correlation as scientists assigned relative dates to the world's rocks according to a proposed sequence of life (fossil evidence).

Radiometric dating provides absolute ages for events in the earth's history. Radiometric dating techniques apply the decay rates of selected naturally radioactive isotopes to stable daughter isotopes to determine how long the unstable parent isotopes have been decaying. Fairly accurate dates have been determined for the events beginning in the Cambrian era; this comprises about 12 percent of the earth's history.

Materials and Equipment

The following materials will be needed by each group of two students:

- A paper strip, such as adding machine tape or shelf paper,
- A meter stick,
- Masking or cellophane tape.

If students use the scale suggested—1 millimeter to 1 million years and 1 meter to 1 billion years—a paper strip 5 meters in length efficiently accommodates the 4.5-billion-year time scale. Make copies of Student Investigation Sheet A on page 91 ("Approximate Ages of Events in Years Before the Present"). Students will use this page to conduct their investigation.

Instructional Strategy

Engage Ask students how long a million years is. How would students count or measure a million of anything? Use this discussion to help students arrive at the question: How does a million years, or even the time since the last ice age, compare

Student Investigation Sheet A

Approximate Ages of Events in Years Before the Present

1. Oldest known rocks and fossils, 3.8 billion years ago.
2. First known plants (algae), 3.2 billion years ago.
3. First known animal (jellyfish), 1.2 billion years ago.
4. Beginning of the Cambrian and first abundant fossils, 550 million years ago.
5. Beginning of the Ordovician and first backboned animals, 500 million years ago.
6. Beginning of the Silurian and the first land plants, 440 million years ago.
7. Beginning of the Devonian and the first amphibians, 400 million years ago.
8. Beginning of the Mississippian, 350 million years ago.
9. Beginning of the Pennsylvanian and the first reptiles, 305 million years ago.
10. Beginning of the Permian, 285 million years ago.
11. Beginning of the Triassic and first dinosaurs, 245 million years ago.
12. Beginning of the Jurassic and first mammals, 205 million years ago.
13. First birds, 150 million years ago.
14. Beginning of the Cretaceous, 140 million years ago.
15. Beginning of the Paleocene and first primates, 65 million years ago.
16. Beginning of the Eocene, 60 million years ago.
17. Beginning of the Oligocene and first elephants, 35 million years ago.
18. Beginning of the Miocene, 25 million years ago.
19. Beginning of the Pliocene, 5 million years ago.
20. First humanlike animals, 2 million years ago.
21. Beginning of the Pleistocene and ice ages, 1 million years ago.
22. Last ice age, 10,000 years ago.

Convert the following to years before the present:

23. Mount Vesuvius eruption destroys Pompeii, A.D. 79.
24. First U.S. satellite orbited, 1958.
25. First man on the moon, 1969.
26. Last New Year's Day.
27. Today.

with the age of the earth? Suppose you want to make a visual model showing a time line of the earth's history, how would you proceed?

Explore Provide students with Student Investigation Sheet A. Have them decide how to represent these events in a time-ordered sequence. Provide a roll of paper tape on which to plot the model.

Students might need help in understanding how to set up a scale that can be displayed in the classroom or adjacent hallway. A reasonable scale is 1 millimeter to 1 million years, 1 centimeter to 10 million years, and 1 meter to 1 billion years. Depending on available space, larger unit distances will be easier to work with. Regardless of the scale the students choose, the last million years will be difficult to plot. Allow students to work out a scale on their own. However, to avoid undue confusion and frustration for some, review student progress after the first few minutes and be ready to ask leading questions or make suggestions.

Allow students time to agree on a reasonable scale, mark the locations of each event on their time scale, and resolve the problem of trying to fit the events from the last 1 million years in the allotted space. When appropriate, encourage students to construct a separate, and larger, scale for marking the most recent events.

Explain Discuss the long period of time in the earth's history before evidence of simple life forms, such as algae, appear in the fossil record. Note that time spans between significant "firsts" become shorter and shorter as you move closer and closer to "today." Compare and discuss expanded scales used to show more detail in the recent past. Discuss the role of scale in helping visualize and better understand the extremely long time span of the geologic time scale and the connections to biological evolution.

Elaborate Challenge students to develop an extended time scale to mark special events in their own lifetime and that of their parents, grandparents, or another adult. Have them calculate the percentage of the earth's history for which there is evidence of life, the percentage of the earth's history for which there is fossil evidence of the first humanlike animals, or the percentage of the earth's history during which dinosaurs lived.

Evaluate Ask students to calculate the length of a paper strip necessary to represent all of geologic time when using the extended scale they used to show the most recent events. Have students write a short news article explaining their scale of geologic time and the evolutionary changes in the earth's lithosphere, atmosphere, and biosphere.

ACTIVITY 7

Proposing the Theory of Biological Evolution: Historical Perspective

This activity uses evolution to introduce students to historical perspectives and the nature of science. The teacher has students read short excerpts of original statements on evolution from Jean Lamarck, Charles Darwin, and Alfred Russel Wallace. This activity is intended as either a supplement to other investigations or as a core activity. Designed for grades 9 through 12, the activity requires a total of three class periods.

Standards-Based Outcomes

The activity provides all students with opportunities to develop understandings of the history and nature of science as described in the *National Science Education Standards*. Specifically, it conveys the following concepts:

- Scientists are influenced by societal, cultural, and personal beliefs and ways of viewing the world. Science is not separate from society but rather a part of society.
- Scientific explanations must meet certain criteria. First and foremost, they must be consistent with experimental and observational evidence about nature, and they must make accurate predictions, when appropriate, about the systems being studied. They should also be logical, respect the rules of evidence, be open to criticism, report methods and procedures, and make knowledge public. Explanations of how the natural world changes based on myths, personal beliefs, religious values, mystical inspiration, superstition, or authority may be personally useful and socially relevant, but they are not scientific.
- Because all scientific ideas depend on experimental and observational confirmation, all scientific knowledge is in principle subject to change as new evidence becomes available. The core ideas of science, such as the conservation of energy or the laws of motion, have been subjected to a wide variety of confirmations and are therefore unlikely to change in the areas in which they have been tested. In areas where data or understanding are incomplete, such as the details of human evolution or questions surrounding global warming, new data may well lead to changes in current ideas or resolve current conflicts. In situations where information is still fragmentary, it is normal for scientific ideas to be incomplete, but this is also where the opportunity for making advances may be greatest.
- Occasionally, there are advances in science and technology that have important and long-lasting effects on science and society.
- The historical perspective of scientific explanations demonstrates how scientific knowledge changes by evolving over time, almost always building on earlier knowledge.

Science Background for Teachers

In historical perspective, explanations for the origin and diversity of life are not new and probably began when humans first began asking questions about the natural world. By the time of the Greeks, individuals such as Thales (624 to 548 B.C.) and Anaximander (611 to 547 B.C.) proposed explanations for life's origins and gradual changes.

In the 1800s three individuals proposed explanations for biological evolution—Jean Lamarck, Charles Darwin, and Alfred Russel Wallace. In the early years of the nineteenth century, a French biologist, Jean Lamarck (1744 to 1829), proposed a view of evolution that questioned the then popular idea that species did not change. Lamarck proposed the idea that changes do take place in animals over long periods of time, specifically through the use of organs and appendages. The popular example of Lamarck's idea is the long necks of giraffes that helped them feed higher in trees. Based on the extension and use of the neck, one generation of giraffes passed the longer neck to the next generation. (See the excerpt for this activity.)

Charles Darwin (1809 to 1882) was born in England and completed his formal education at Cambridge University. Darwin's main interests centered on the study of nature and collecting a diversity of organisms. After graduation, Darwin's professor recommended him for the position of naturalist on H.M.S. *Beagle*. The voyage of the *Beagle* lasted five years (1831 to 1836) and provided the observations and evidence (in the form of specimens) that became the foundation for Darwin's theories. Of particular note in history is Darwin's observations on the Galapagos Islands located off the coast of Ecuador. Darwin's curiosity and insight led him to observe both similarities and differences among organisms and compare them on the mainland and the islands 600 miles offshore. Based on his observations, he wondered about the origin of different plants and animals, and the variations in species he recorded in similar organisms.

After returning to England, Darwin spent more than twenty years studying the specimens, experimenting, and reviewing the notes of his voyage. In 1858 he was surprised to find that Alfred Russel Wallace had formulated similar conclusions. In the same year, Darwin reported his and Wallace's work in a joint presentation to the Scientific Society in London. One year later, in 1859, Darwin published *On the Origin of Species by Means of Natural Selection*. This publication caused great debate and what is now viewed as a scientific revolution. Darwin's theories of evolution have also had considerable impact on society and our cultural views.

Alfred Russel Wallace (1823 to 1913) was also born in England. He became a teacher of English. He later developed an interest in collecting plants and insects. In 1848 he made an expedition to the Amazon River in Brazil to collect scientific materials. On a later expedition to the Malay Islands, Wallace observed some variations in organisms that engaged the same questions that Darwin posed—why did each island have different species? Wallace thought about the question for three years and in 1858 he proposed his theory.

Materials and Equipment

Excerpt from *Zoological Philosophy* by Jean Lamarck (provided)

Excerpt from *On the Tendency of Varieties to Depart Indefinitely from the Original Type* by Alfred Russel Wallace (provided)

Excerpt from *On the Origin of Species* by Charles Darwin (provided)

Instructional Strategy

These excerpts give the students an opportunity to read original statements by individuals who contributed to a major revolution in the history of biology. The instructional strategy is that of small-group discussions. Students read an original excerpt prior to class and discuss the reading in class.

Engage Introduce the sequence of readings by asking questions based on the learning outcomes:

- How do you think the society in which scientists live might influence their views?
- What makes a person's explanation scientific?
- Can scientific explanations change? If so, how? Why? If not, why not?
- Can you name some major theories in science? In biology?

Ask the students what they know about the theory of evolution. What do they know about Charles Darwin? When did he propose his theory? Did any other individuals propose theories about evolution? How did Darwin develop his theory of evolution? Questions such as these will set the stage for the first reading. Assign the reading by Jean Lamarck as homework.

Explore Students should work in groups of four to discuss Jean Lamarck's explanations of changes in organisms. Questions for student discussions include:

- What is the role of the environment in Lamarck's explanation?
- What scientific approach is suggested by Lamarck's statement: "Nothing of all this can be considered as hypothesis or private opinion; on the contrary, they are truths which, in order to be made clear, only require attention and the observation of facts."
- Was Lamarck's explanation scientific? Why or why not?
- Can you propose any other explanations for Lamarck's observations about the disuse and use of organs?

Explain Prior to this group discussion, assign the reading by Alfred Russel Wallace. With your guidance, this discussion should clarify for students some of the fundamental concepts about science as a human endeavor and the nature of science. This should include discussion in groups of four followed by a full class summary of the learning outcomes.

- How would you characterize Wallace's idea that "The life of wild animals is a struggle for existence?" How is Wallace's view scientific?
- Wallace claims that "useful variations will tend to increase, unuseful or hurtful variations to diminish." How does this occur? What evidence does he cite?
- How does Wallace's explanation differ from Lamarck's?
- What do you think of Wallace's critique of Lamarck's hypotheses?

Elaborate Prior to this group discussion, assign the reading by Charles Darwin. In these discussions, students should apply concepts about the nature of science and the historical perspective developed during prior discussions. This discussion should demonstrate greater sophistication and understanding by the students.

- What led Darwin to formulate his ideas about the origin of species?
- On what did he base his explanations?
- What did Darwin propose as the origin of species?
- What was the relationship of Lamarck's and Wallace's work to Darwin's?
- Was Darwin's explanation scientific? Why or why not?
- How did Darwin attempt to determine how modifications of a species are accomplished?
- How did Darwin explain the incomplete nature of his ideas?

Evaluate Have each student write a brief essay on the nature of scientific knowledge as demonstrated in the development of the theory of evolution. They should cite at least two quotes from the reading to support their discussion. The essays should incorporate the concepts of adaptation, natural selection, and descent from common ancestors.

Student Sheet

Zoological Philosophy

Jean Lamarck (1809)

The environment affects the shape and organization of animals, that is to say that when the environment becomes very different, it produces in course of time corresponding modifications in the shape and organization of animals.

If a new environment, which has become permanent for some race of animals, induces new habits in these animals, that is to say, leads them into new activities which become habitual, the result will be the use of some one part in preference to some other part, and in some cases the total disuse of some part no longer necessary.

Nothing of all this can be considered as hypothesis or private opinion; on the contrary, they are truths which, in order to be made clear, only require attention and the observation of facts.

Snakes have adopted the habit of crawling on the ground and hiding in the grass; so that their body, as a result of continually repeated efforts at elongation for the purpose of passing through narrow spaces, has acquired a considerable length, quite out of proportion to its size. Now, legs would have been quite useless to these animals and consequently unused. Long legs would have interfered with their need of crawling, and very short legs would have been incapable of moving their body, since they could only have had four. The disuse of these parts thus became permanent in the various races of these animals, and resulted in the complete disappearance of these same parts, although legs really belong to the plan or organization of the animals of this class.

The frequent use of any organ, when confirmed by habit, increases the functions of that organ, leads to its development, and endows it with a size and power that it does not possess in animals which exercise it less.

We have seen that the disuse of any organ modifies, reduces, and finally extinguishes it. I shall now prove that the constant use of any organ, accompanied by efforts to get the most out of it, strengthens and enlarges that organ, or creates new ones to carry on the functions that have become necessary.

The bird which is drawn to the water by its need of finding there the prey on which it lives, separates the digits of its feet in trying to strike the water and move about on the surface. The skin which unites these digits at their base acquires the habit of being stretched by these continually repeated separations of the digits; thus in course of time there are formed large webs which unite the digits of ducks, geese, etc. as we actually find them.

It is interesting to observe the result of habit in the peculiar shape and size of the giraffe; this animal, the largest of the mammals, is known to live in the interior of Africa in places where the soil is nearly always arid and barren, so that it is obliged to browse on the leaves of trees and to make constant efforts to reach them. From this habit long maintained in all its race, it has resulted that the animal's fore-legs have become longer than its hind legs, and that its neck is lengthened to such a degree that the giraffe, without standing up on its hind legs, attains a height of six metres (nearly twenty feet).

Philosophie Zoologique. Paris. 1809. Translated by H. Elliott, Macmillan Company, London. 1914.

Student Sheet

On the Tendency of Varieties to Depart Indefinitely from the Original Type

Alfred Russel Wallace (1858)

The Struggle for Existence

The life of wild animals is a struggle for existence. The full exertion of all their faculties and all their energies is required to preserve their own existence and provide for that of their infant offspring. The possibility of procuring food during the least favorable seasons and of escaping the attacks of their most dangerous enemies are the primary conditions which determine the existence both of individuals and of entire species.

The numbers that die annually must be immense; and as the individual existence of each animal depends upon itself, those that die must be the weakest—the very young, the aged, and the diseased—while those that prolong their existence can only be the most perfect in health and vigor, those who are best able to obtain food regularly and avoid their numerous enemies. It is "a struggle for existence," in which the weakest and least perfectly organized must always succumb.

Useful Variations Will Tend to Increase, Unuseful or Hurtful Variations to Diminish

Most or perhaps all the variations from the typical form of a species must have some definite effect, however slight, on the habits or capacities of the individuals. Even a change of color might, by rendering them more or less distinguishable, affect their safety; a greater or less development of hair might modify their habits. More important changes, such as an increase in the power or dimensions of the limbs or any of the external organs, would more or less affect their mode of procuring food or the range of country which they could inhabit. It is also evident that most changes would affect, either favorable or adversely, the powers of prolonging existence. An antelope with shorter or weaker legs must necessarily suffer more from the attacks of the feline carnivora; the passenger pigeon with less powerful wings would sooner or later be affected in its powers of procuring a regular supply of food; and in both cases the result must necessarily be a diminution of the population of the modified species.

If, on the other hand, any species should produce a variety having slightly increased powers of preserving existence, that variety must inevitably in time acquire a superiority in numbers.

Lamarck's Hypothesis Very Different from that Now Advanced

The hypothesis of Lamarck—that progressive changes in species have been produced by the attempts of animals to increase the development of their own organs and thus modify their structure and habits—has been repeatedly and easily refuted by all writers on the subject of varieties and species.

The giraffe did not acquire its long neck by desiring to reach the foliage of the more lofty shrubs and constantly stretching its neck for the purpose, but because any varieties which occurred among its ancestors with a longer neck than usual at once secured a fresh range of pasture over the same ground as their shorter-necked companions, and on the first scarcity of food were thereby enabled to outlive them.

Journal of the Proceedings of the Linnean Society
August 1858, London

Student Sheet

On the Origin of Species

Charles Darwin (1859)

Introduction

When on board H.M.S. *Beagle*, as naturalist, I was much struck with certain facts in the distribution of the inhabitants of South America, and in the geological relations of the present to the past inhabitants of that continent. These facts seemed to me to throw some light on the origin of species—that mystery of mysteries, as it has been called by one of our greatest philosophers. On my return home, it occurred to me, in 1837, that something might perhaps be made out on this question by patiently accumulating and reflecting on all sorts of facts which could possibly have any bearing on it. After five years work I allowed myself to speculate on the subject, and drew up some short notes; these I enlarged in 1844 into a sketch of the conclusions, which then seemed to me probable; from that period to the present day I have steadily pursued the same object. I hope that I may be excused for entering on these personal details, as I give them to show that I have not been hasty in coming to a decision.

My work is now nearly finished; but as it will take me two or three more years to complete it, and as my health is far from strong, I have been urged to publish this Abstract. I have more especially been induced to do this, as Mr. Wallace, who is now studying the natural history of the Malay archipelago, has arrived at almost exactly the same general conclusions that I have on the origin of species. Last year he sent to me a memoir on this subject, with a request that I would forward it to Sir Charles Lyell, who sent it to the Linnean Society, and it is published in the third volume of the Journal of that Society. Sir C. Lyell and Dr. Hooker, who both knew of my work—the latter having read my sketch of 1844—honoured me by thinking it advisable to publish, with Mr. Wallace's excellent memoir, some brief extracts from my manuscripts.

In considering the Origin of Species, it is quite conceivable that a naturalist, reflecting on the mutual affinities of organic beings, on their embryological relations, their geographical distribution, geological succession, and other such facts, might come to the conclusion that each species had not been independently created, but had descended, like varieties, from other species. Nevertheless, such a conclusion, even if well founded, would be unsatisfactory, until it could be shown how the innumerable species inhabiting this world have been modified, so as to acquire that perfection of structure and coadaptation which most justly excites our admiration. Naturalists continually refer to external conditions, such as climate, food, etc., as the only possible cause of variation. In one very limited sense, as we shall hereafter see, this may be true; but it is preposterous to attribute to mere external conditions, the structure, for instance, of the woodpecker, with its feet, tail, beak, and tongue, so admirable adapted to catch insects under the bark of trees. In the case of the misseltoe, which draws its nourishment from certain trees, which has seeds that must be transported by certain birds, and which has flowers with separate sexes absolutely requiring the agency of certain insects to bring pollen from one flower to the other, it is equally preposterous to account for the structure of this parasite, with its relations to several distinct organic beings, by the effects of external conditions, or of habit, or of the volition of the plant itself.

The author of the 'Vestiges of Creation' would, I presume, say that, after a certain unknown number of generations, some bird had given birth to a woodpecker, and some plant to the misseltoe, and that these had been produced perfect as we

(Continued on page 99)

Student Sheet

(Continued from page 98)

now see them; but this assumption seems to me to be no explanation, for it leaves the case of the coadaptations of organic beings to each other and to their physical condition of life, untouched and unexplained.

It is, therefore, of the highest importance to gain a clear insight into the means of modification and coadaptation. At the commencement of my observations it seemed to me probable that a careful study of domesticated animals and of cultivated plants would offer the best chance of making out this obscure problem. Nor have I been disappointed; in this and in all other perplexing cases I have invariable found that our knowledge, imperfect though it be, of variation under domestication, afforded the best and safest clue. I may venture to express my conviction of the high value of such studies, although they have been very commonly neglected by naturalists.

No one ought to feel surprise at much remaining as yet unexplained in regard to the origin of species and varieties, if he makes due allowance for our profound ignorance in regard to the mutual relations of all the beings which live around us. Who can explain why one species ranges widely and is very numerous, and why another allied species has a narrow range and is rare? Yet these relations are of the highest importance, for they determine the present welfare, and, as I believe, the future success and modification of every inhabitant of this world. Still less do we know of the mutual relations of the innumerable inhabitants of the world during the many past geological epochs in its history. Although much remains obscure, and will long remain obscure, I can entertain no doubt, after the most deliberate study and dispassionate judgment of which I am capable, that the view which most naturalists entertain, and which I formerly entertained—namely, that each species has been independently created—is erroneous. I am fully convinced that species are not immutable; but that those belonging to what are called the same genera are lineal descendants of some other and generally extinct species, in the same manner as the acknowledged varieties of any one species are the descendants of that species. Furthermore, I am convinced that Natural Selection has been the main but not exclusive means of modification.

On the Origin of Species by Means of Natural Selection.
London. 1859.

ACTIVITY 8

Connecting Population Growth and Biological Evolution

In this activity, students develop a model of the mathematical nature of population growth. The investigation provides an excellent opportunity for consideration of the population growth of plant and animal species and the resultant stresses that contribute to natural selection. This activity will require two class periods and is appropriate for grades 5 through 12. The activity is based on an original activity from the Earth Science Curriculum Project. It is used with permission.[14]

Standards-Based Outcomes

This activity provides all students an opportunity to develop understandings about scientific inquiry and biological evolution as described in the *National Science Education Standards.* Specifically, it conveys the following concepts:

• Mathematics is essential in scientific inquiry. Mathematical tools and models guide and improve the posing of questions, gathering data, constructing explanations, and communicating results.

• Species evolve over time. Evolution is the consequence of (1) the potential for a species to increase its numbers, (2) the genetic variability of offspring due to mutation and recombination of genes, (3) a finite supply of the resources required for life, and (4) the ensuing selection of those offspring better able to survive and leave offspring in a particular environment. (Item 1 is the primary content emphasis of this activity. Teachers can introduce the other factors as appropriate.)

• Populations grow or decline through the combined effects of births and deaths and through emigration and immigration into specific areas. Populations can increase through linear or exponential growth, with effects on resource use and on environmental pollution.

• Populations can reach limits to growth. Carrying capacity is the maximum number of organisms that can be supported by a given environment.

• Living organisms have the capacity to produce populations of arbitrarily large size, but environments and resources are finite. This fundamental tension has profound effects on the interactions between organisms.

Science Background for Teachers

The tension between expanding populations and limited resources was a fundamental point that Darwin came to understand when he read Thomas Malthus.[15] This understanding subsequently had an important influence on the formulation of his theory of natural selection.

This activity extends the general idea of population growth to humans. Here the important point is that human beings live within the world's ecosystems. Increasingly, humans modify ecosystems as a result of population growth, technology, and consumption. Human destruction of habitats through direct harvesting, pollution, atmospheric changes, and other factors is threatening current global stability, and, if not addressed, ecosystems will be irreversibly affected.

The increase in the size of a population (such as the human population) is an example of exponential growth. The human population grew at the slow rate of only about 0.002 percent a year for the first several million years of our existence. Since then the average annual rate of human population has increased to an all-time high of 2.06 percent in 1970. As the base number of people undergoing growth has increased, it has taken less and less time to add each new billion people. It took 2 million years to add the first billion people; 130 years to add the second billion; 30 years to add the third billion; 15 years to add the fourth billion; and only 12 years to add the fifth billion. We are now approaching the sixth billion.

Materials and Equipment

Each group of three or four students will need:

- Approximately 2,000 small, uniformly shaped objects (kernels of corn, dried beans, wooden markers, plastic beads)
- 10 paper cups or small beakers
- A 250-ml or 400-ml beaker

Instructional Strategy

Engage Initiate a discussion on human population with such questions as: How long have humans been on the earth? How do you think the early rate of human population growth compares with the population growth rate today? Why did this rate change?

Tell students that this investigation represents a model of population growth rates.

Explore Have student groups complete the following activities.

- Place the glass beakers on their desks. Begin by placing two objects (e.g., corn or plastic beads) in it. The beaker represents the limits of an ecosystem or ultimately the earth.
- Place 10 cups in a row on their desk. In the first cup, place two objects. In the second cup, place twice as many objects as the first cup (four). Have students record the number of objects on the outside of the cup. Continue this procedure by placing twice as many objects as in the former cup, or doubling the number, in cups 3 through 10. Be sure students record the numbers on the cups.
- Take the beaker and determine its height. Have students indicate the approximate percentage of volume that is *without* objects. Record this on the table as 0 time.
- At timed intervals of 30 seconds, add the contents of cups 1 through 10. Students should record the total population and approximate percentage of volume in the beaker that is without objects.
- Students should complete the procedure and graph their results as total population versus results.

Students may question the need for the 30-second intervals. The length of the time interval is arbitrary. Any time interval will do. Preparation of the graph can be assigned as homework.

Range of Results

The mathematics involved in answering the questions may challenge some students. Assist students when necessary to enable them to accomplish the objectives of the investigation. Table 1 shows the population and the percent of the beaker's volume without objects. A typical student graph is shown in Figure 1.

Explain Ask the students to explain the relationship between population growth and biological evolution in populations of microorganisms, plants, and animals. Through questions and discussion, help them develop the connections stated in the learning outcome for the activity. Evolution results from an interaction of factors related to the potential for species to increase in numbers, the genetic variability in a population, the supply of essential resources, and environmental pressures for selection of those offspring that are able to survive and reproduce.

Elaborate Begin by having students explain the results of their activity. During the discussion of the graph, have the students consider some of the following: Are there any limitations to the number of people the earth will support? Which factor might limit population growth first? How does this factor relate to human evolution? Are

Table 1
Population growth

Time Internal	Population	Percentage of empty volume (400-ml beaker)
0	2	99%
1	4	99%
2	8	99%
3	16	98%
4	32	97%
5	64	95%
6	128	93%
7	256	88%
8	512	80%
9	1024	70%
10	2048	50%
11	4096	0%

Figure 1
Sample population growth graph

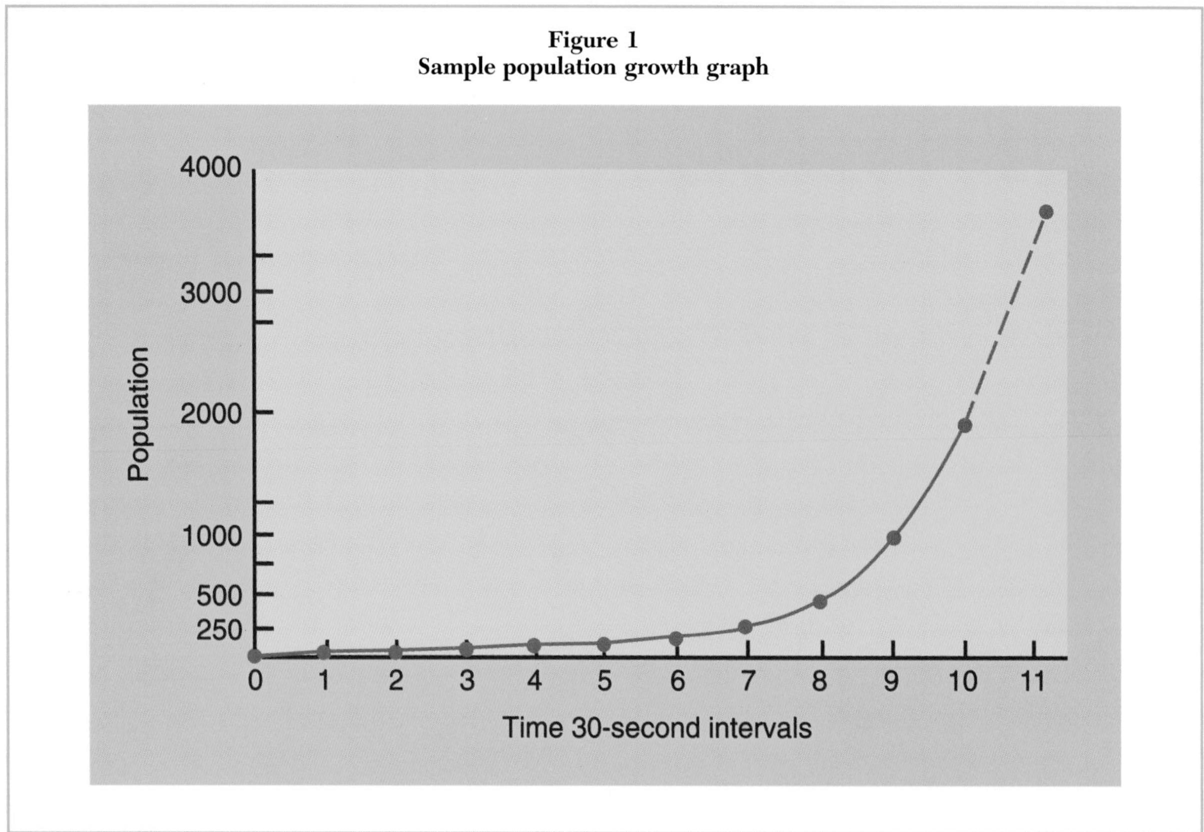

there areas in the world where these limits have been reached already? Have we gone beyond the earth's ideal population yet? What problems will we face if we overpopulate the earth? How might human influence on, for example, habitats affect biological evolution. Students' answers to these questions will vary, depending on their background and information. The outcome, however, should be an intense discussion of some vital problems and should provide opportunities to introduce the fundamental concepts from the *National Science Education Standards*.

Evaluation 1. Human population on the earth is thought to have had a slow start, with doubling periods as long as 1 million years. The current world population is thought to be doubling every 37 years. How would this growth rate compare with the rates found in your investigation?

Both the population in the investigation and on the earth increase in a geometric progression. This means the graphs have the same shape. You can substitute 37 years for every 30-second interval and the numbers will represent actual world population growth. The slope of the graph would remain the same.

2. What happens to populations when they reach the limits to growth?

The populations stop growing because death rates (or emigration) exceed birth rates (or immigration).

NOTES

1. National Research Council. 1996. *National Science Education Standards*. Washington, DC: National Academy Press. www.nap.edu/readingroom/books/nses
2. A Draft Growth-of-Understanding Map derived from *Benchmarks for Science Literacy* (Jan. 1998), AAAS (American Association for the Advancement of Science) Project 2061.
3. Biological Sciences Curriculum Study (BSCS). 1978. *Biology Teachers' Handbook*. 3rd ed. William V. Mayer, ed. New York: John Wiley and Sons, pp. 350-352.
4. *Standards,* p. 117.
5. Jonathan Weiner. 1994. *The Beak of the Finch: A Story of Evolution in Our Time.* New York: Alfred A. Knopf.
6. Tijs Goldschmidt. 1996. *Darwin's Dreampond: Drama in Lake Victoria*. Cambridge, MA: MIT Press.
7. *BSCS Biology: A Human Approach*. 1997. Dubuque, IA: Kendall/Hunt Publishing Co., pp. 47-49 and pp. 64-69.
8. See Chapter 2 of this document for more discussion on genetic variation and natural selection, and pages 158 and 185 of the *National Science Education Standards*.
9. *Evolution: Inquiries into Biology and Earth Science by BSCS*. 1992. Seattle: Videodiscovery, pp. 49-53 and pp. 211-221.
10. *Standards,* p. 117.
11. Earth Science Curriculum Project (ESCP). 1973. *Investigating the Earth*. rev. ed. Boston, MA: Houghton Mifflin.
12. Please review pages 143-148 of the *National Science Education Standards*.
13. *Investigating the Earth*.
14. *Investigating the Earth*.
15. Thomas Malthus. 1993. *Essay on the Principle of Population*. Geoffrey Gilbert, ed. Oxford: Oxford University Press.

7

Selecting Instructional Materials

Quality instructional materials are essential in teaching about evolution and the nature of science.

It also is important to consider the context within which specific materials will be used. This chapter therefore begins with brief discussions of school science programs and the criteria used to design curricula.

Criteria for Contemporary Science Curriculum

Before selecting specific materials to teach evolution and the nature of science, it is important to identify criteria that can help evaluate school science programs and the design of instructional materials. Chapter seven in the *National Science Education Standards*, "Science Education Program Standards," describes the conditions needed for quality school science programs. These conditions focus on six areas:

- Consistency across all elements of the science program and across the K-12 continuum
- Quality in the program of studies
- Coordination with mathematics
- Quality resources
- Equitable opportunities for achievement
- Collaboration within the school community to support a quality program

Similarly, educators need to consider criteria against which to judge instructional materials. Teachers, curriculum designers, and other school personnel can use the following criteria to evaluate the design of a new curriculum, to select instructional materials, or to adapt instructional materials through professional development. No set of instructional materials will meet all the following criteria. You will have to make a judgment about the degree to which materials meet criteria and about acceptable and unacceptable omissions. These criteria are adapted from earlier discussions of standards-based curriculum.[1]

Criterion 1: *A Coherent, Consistent, and Coordinated Framework for Science Content.* Science content should be consistent with national, state, and local standards and benchmarks. Whether for lessons, units, or a complete elementary, middle, or high school program, the content should be well-thought-out, coordinated, and conceptually, procedurally, and coherently organized. The roles of science concepts, inquiry, science in personal and social contexts, and the history and nature of science should be clear and explicit.

Criterion 2: *An Organized and Systematic Approach to Instruction.* Most contemporary science curricula incorporate an instructional model. The instructional model should (1) provide for different forms of interaction among students and between the teachers and students, (2) incorporate a variety of teaching strategies, such as inquiry-oriented investigations, cooperative groups, use of technology, and (3) allow adequate time and opportunities for students to acquire knowledge, skills, and attitudes.

Criterion 3: *An Integration of Psychological Principles Relative to Cognition, Motivation, Development, and Social Psychology.* Psychological principles such as those found in the American Psychological Association publication *How Students Learn: Reforming School Through Learner-Centered Education*[2] should be applied to the framework for content, teaching, and assessment. These psychological principles include more than learning theory. They include providing for motivation, development, and social interactions.

Criterion 4: *Varied Curriculum Emphases.* The idea of curriculum emphases can be expressed by thinking about the foreground and background in a painting. An artist decides what will be in the foreground, and that subject is emphasized. Science curricula can, for example, emphasize science concepts, inquiry, or the history and nature of science, while other goals may be evident but not emphasized. No one curriculum emphasis is best for all students; probably, a variety of emphases accommodates the interests, strengths, and demands of science content.

Criterion 5: *An Array of Opportunities to Develop Knowledge, Understanding, and Abilities Associated with Different Dimensions of Scientific Literacy.* Contemporary science curricula should provide a balance among the different dimensions of science literacy, which include an understanding of scientific concepts, the ability to engage in inquiry, and a capacity to apply scientific information in making decisions.[3]

Criterion 6: *Teaching Methods and Assessment Strategies Consistent with the Goal of Science Literacy.* Approaches to teaching and assessment ought to be consistent with the goals of teaching evolution, inquiry, and the history and nature of science. This can be accomplished by using inquiry-oriented teaching methods and by assessing students during investigative activities.

Criterion 7: *Professional Development for Science Teachers Who Implement the Curriculum.* Curricula need to provide opportunities that support teachers as they develop the knowledge and skills associated with implementing and institutionalizing the science program.

Criterion 8: *An Inclusion of Appropriate Educational Technologies.* The use of computers and various types of software enhances learning when students use the technologies in meaningful ways. The use of educational technologies should be consistent with other features of the curriculum—for instance, the dimensions of scientific literacy and an instructional model.

Criterion 9: *Thorough Field Testing and Review for Scientific Accuracy and Pedagogic Quality.* One important legacy of the 1960s curriculum reform is the field testing of materials in a variety of science classrooms. Field testing and reviewing a program identify problems that developers did not recognize and fine tune the materials to the varied needs of teachers, learners, and schools. Scientists should review materials for accuracy. Developers can miss the subtleties of scientific concepts, inquiry, and design. In addition, educators who review materials can provide valuable insights about teaching and assessment that help developers improve materials and enhance learning.

Criterion 10: *Support from the Educational System.* Research on the adoption, implementation, and change associated with curricula indicates the importance of intellectual, financial, and moral support from those within the larger educational system.[4] This support includes science teachers, administrators, school boards, and communities. Although a curriculum cannot ensure support, it should address the need for support and provide indicators of support, such as provision of materials and equipment for laboratory investigations, budget allocations for professional development, and proclamations by the school board.

Clearly, no one curriculum thoroughly incorporates all ten criteria. There are always trade-offs when developing, adapting, or adopting a science curriculum. However, the criteria should provide assistance to those who have the responsibility of improving the science curriculum.

Analyzing Instructional Materials

The process of selecting quality materials includes determining the degree to which they are consistent with the goals, principles, and criteria developed in the *National Science Education Standards*. Well-defined selection criteria help ensure a thoughtful and effective process. To be both usable and defensible, the selection criteria must be few in number and embody the critical tenets of accurate science content, effective teaching strategies, and appropriate assessment techniques.

The process described in the following pages can help teachers, curriculum designers, or other school personnel complete a thorough and accurate evaluation of instructional materials. To help make this examination both thorough and usable, references to specific pages and sections in the *National Science Education Standards* have been provided, as have worksheets to keep track of the information needed to analyze and select the best instructional materials.

Analysis Procedures

The procedures outlined in this section include:

- Overview of instructional materials
- Analysis of science subject matter
- Analysis of pedagogy
- Analysis of assessment process
- Evaluating the teacher's guide
- Analysis of use and management

The extent to which instructional materials meet the criteria outlined in this chapter determines their usefulness for classroom teachers and the degree of alignment with the *Standards*. A thorough analysis of instructional materials requires considerable time and collaboration with others and attention to detail. Good working notes are helpful in this process. We recommend using the analysis worksheets provided at the end of this chapter.

Overview of Instructional Materials

The following overview of instructional materials introduces the review process and provides a general context for analysis and subsequent selection of specific materials.

1. The first consideration is whether the key concepts of evolution and the nature of science are being emphasized. To help make this determination, locate the table of contents, index, and glossary in the material you are evaluating. The box below contains terms related to fundamental concepts in evolution and the nature of science taken from the *Standards*. Record page numbers where each is found for future reference. (See Worksheet 1 on page 112 in the back of this chapter.) These terms will give you a preliminary indication of coverage on these fundamental topics.

Evolution

evolution, diversity, adaptation, interpreting fossil evidence, techniques for age determination, natural selection, descent from common ancestors

Nature of Science

explanation, experiment, evidence, inquiry, model, theory, skepticism

2. Look through both student and teacher materials. Are student outcomes listed? Note page numbers for several outcomes related to evolution and the nature of science.

3. Look for student investigations or activities. Where are they located? Note that in some materials, student investigations are integrated within the reading material. In others they are located in a separate section—sometimes at the back of a chapter or book or in a separate laboratory manual.

4. Read several relevant paragraphs of student text material. What is your judgment about the concepts? Are the concepts in the students' text

consistent with the fundamental concepts in the *Standards*? Does the text include more, fewer, or different concepts?

5. Do the photographs and illustrations provide further understanding of the fundamental concepts?

Analysis of Instructional Materials for Science Subject Matter

A. CONTENT

The following procedures for content analysis will help you examine instructional materials for fundamental concepts of evolution, science as inquiry, and the nature of science. Look for evidence in discussions in the text and in the student investigations to determine the degree to which the fundamental concepts are addressed. Fundamental concepts underlying specific standards on evolution and the nature of science are referenced below. (Note: You will need a copy of the *National Science Education Standards* or access to it through the World Wide Web at www.nap.edu/readingroom/books/nses.)

Content Standard C—Life Science: *grades 5-8*, "Diversity and Adaptations of Organisms," p. 158; *grades 9-12*, "Biological Evolution," p. 185; also read "Developing Student Understanding" *grades 5-8*, pp. 155-156; and *grades 9-12*, p. 181.

Content Standard D—Earth and Space Science: *grades 5-8*, "Earth's History," p. 160; *grades 9-12*, "The Origin and Evolution of the Earth System," pp. 189-190; also read "Developing Student Understanding," *grades 5-8*, pp. 158-159; *grades 9-12*, pp. 187-188.

1. Choose a lesson or representative section of the student instructional materials on the topic of evolution. Make a preliminary list of the fundamental concepts from the *Standards* that are included in the lesson and place them on your worksheet. (See Worksheet 2 on page 114 in the back of this chapter.)

2. Select one of these fundamental concepts and list all sections of the materials that deal with this idea. Determine whether the materials focus on the fundamental concepts, or if they represent only a superficial match. For example, Life Science Standard C in the *Standards*[5] specifies: "Biological evolution accounts for the diversity of species developed through gradual processes over many generations. Species acquire many of their unique characteristics through biological adaptation, which involves the selection of naturally occurring variations in populations." The instructional materials should provide opportunities for students to develop an understanding of biodiversity and evolution as described in the *Standards*. A negative example would be defining the term biodiversity only in reference to the fact that wide varieties of plants and animals populate particular environments.

You should complete this analysis for all fundamental concepts associated with a particular standard. The more fundamental concepts you analyze using this process, the more confidence you will have in the quality of the instructional materials and their alignment with the *Standards*. Identify the fundamental concepts that are not developed and the variation of treatment among those that are included in the materials.

3. If appropriate, select one of the student investigations for analysis of subject matter. On what fundamental concepts from Life Science Standard C or Earth and Space Science Standard D is the investigation focused? To what degree does the activity fulfill the intent of the fundamental concepts? For example, making and comparing model casts and molds of sea shells does not necessarily contribute to an understanding of how fossils are formed or provide important evidence of how life and environmental conditions have changed. It is recommended that you analyze a second student investigation.

B. SCIENTIFIC INQUIRY

1. You should develop some understanding of scientific inquiry in the *Standards*. Read Standard A, Science as Inquiry, referenced on the following page.

Standard A—Science as Inquiry: *grades* 5-8, pp. 145-148; *grades 9-12*, pp. 175-176; also read "Developing Student Understanding," *grades* 5-8, pp. 143-144; *grades 9-12*, pp. 173-174.

Note that Standard A specifies two separate aspects of science as inquiry: abilities necessary to do scientific inquiry, and fundamental understandings about scientific inquiry. Examine several lessons in the student and teacher materials to answer the following question: To what degree do the lessons provide students the opportunity to develop the abilities and understandings of scientific inquiry?

2. Read through the text narrative, looking for student investigations and examining any suggestions for activities outside of class time. Are opportunities provided for students to develop abilities of scientific inquiry such as posing their own relevant questions, planning and conducting investigations, using appropriate tools and techniques to gather data, using evidence to communicate defensible explanations of cause and effect relationships, or using scientific criteria to analyze alternative explanations to determine a preferred explanation? Record page numbers where examples are found and make notes of explanation.

2. What opportunities are provided for students to develop a fundamental understanding of scientific inquiry? In addition to the language of the text, examine the teacher's guide for suggestions that teachers can use to discuss the role and limitations of scientific skills such as making observations, organizing and interpreting data, and constructing defensible explanations based on evidence. Can you find a discussion of how science advances through legitimate skepticism? Can you find a discussion of how scientists evaluate proposed explanations of others by examining and comparing evidence, identifying reasoning that goes beyond the evidence, and suggesting alternative explanations for the same evidence? Are there opportunities for students to demonstrate these same understandings as a part of their investigations? Make notes where this evidence is found for later reference.

C. HISTORY AND NATURE OF SCIENCE

1. Are history and the nature of science incorporated into the treatment of evolution? Read Standard G, History and Nature of Science, referenced in the following box.

Content Standard G—History and Nature of Science: *grades* 5-8, pp. 170-171; *grades* 9-12, pp. 200-201 and p. 204; also read "Developing Student Understanding," *grades* 5-8, p. 170; *grades* 9-12, p. 200.

2. Read through several lessons in the student and teacher materials. Can you find examples describing the roles of scientists, human insight, and scientific reasoning in the historical and contemporary development of explanations for evolution? Can you find specific references to historical contributions of scientists in the development of fundamental concepts of evolution? What evidence can you find in the text narrative or student investigations that demonstrates how scientific explanations are developed, reviewed by peers, and revised in light of new evidence and thinking?

Analysis of Pedagogy

What students learn about evolution and the nature of science depends on many things, including the accuracy and developmental appropriateness of content and its congruence with the full intent of the content standards. Opportunities to learn should be consistent with contemporary models of learning. The criteria in this section are based on characteristics of effective teaching proposed in Teaching Standards A, B, and E.

Teaching Standard A—Teachers of science plan an inquiry-based science program for their students, pp. 30-32.

Teaching Standard B—Teachers of science guide and facilitate learning, pp. 32-33 and 36-37.

Teaching Standard E—Teachers of science develop communities of science learners that reflect the intellectual rigor of scientific inquiry and the attitudes and social values conducive to science learning, pp. 45-46 and 50-51.

Using the following sequence of questions, examine several lessons in the student materials and the teacher's guide. (See Worksheet 3 on page 117 in the back of this chapter.)

1. Do the materials identify specific learning goals or outcomes for students that focus on one or more of the fundamental concepts of evolution and the nature of science?

2. Study the opening pages of a relevant chapter or section. Does the material on the opening pages of the chapter or section on evolution engage and focus student thinking on interesting questions, problems, or relevant issues?

3. Does the material provide a sequence of learning activities connected in such a way as to help students build understanding of a fundamental concept? Are suggestions provided to help the teacher keep students focused on the purpose of the lesson?

4. Does the teacher's guide present common student misconceptions related to the fundamental concepts of evolution and the nature of science? Are suggestions provided for teachers to find out what their students already know? Are there learning activities designed to help students confront their misconceptions and encourage conceptual change?

Analysis of Assessment Process

Assessment criteria in this section are grounded in the Assessment Standards.

Assessment Standards A to E, Chapter 5, pp. 78-87.

Examine several lessons in the student and teacher materials for evidence to answer the following questions. (See Worksheet 4 on page 118 in the back of this chapter.)

1. Is there consistency between learning goals and assessment? For example, if instruction focuses on building understanding of fundamental concepts, do assessments focus on explanations and not on vocabulary?

2. Do assessments stress application of concepts to new or different situations? For example, are the students asked to explain new situations with concepts they have learned?

3. Are assessment tasks fair for all students? For example, does success on assessment tasks depend too heavily on the student's ability to read complex items or write explanations as opposed to understanding the fundamental concepts?

4. Are suggestions for scoring criteria or rubrics provided for the teacher?

Evaluating the Teacher's Guide

Examine several lessons in the teacher's guide to help answer the following questions:

1. Does the teacher's guide present appropriate and sufficient background on science?

2. Are the suggested teaching strategies usable by most teachers?

3. Are suggestions provided for pre- and post-investigation discussions focusing on concept development, inquiry, and the nature of science?

4. Does the teacher's guide recommend additional professional development?

5. Does the teacher's guide indicate the types of support teachers will need for the instructional materials?

Analysis of Use and Management

A high degree of alignment with *Standards* content, pedagogy, and assessment criteria does not necessarily guarantee that instructional materials will be easy to manage. The *Standards* address the importance of professional development, and some aspects of the program standards apply as well.[6]

1. How many different types of materials must be managed and orchestrated during a typical chapter, unit, or teaching sequence (e.g., student text, teacher's guide, transparencies, handouts,

videos, and software)? (See Worksheet 5 on page 119 in the back of this chapter.)

2. Does the teacher's guide contain suggestions for effectively managing materials?

3. Do the instructional materials call for equipment, supplies, and technology that teachers may not have?

4. Do the instructional materials identify safety issues and provide adequate precautions?

5. Is the cost for materials and replacements reasonable? Are there special requirements?

NOTES

1. Rodger Bybee. 1997. *Achieving Scientific Literacy: From Purposes to Practices*. Portsmouth, NH: Heinemann. Rodger Bybee, 1996. *National Standards and the Science Curriculum*. Dubuque, IA: Kendall/Hunt Publishing Co.
2. N. M. Lambert and B. L. McCombs. 1998. *How Students Learn: Reforming Schools Through Learner-Centered Education*. Washington, DC: American Psychological Association.
3. National Research Council. 1996. *National Science Education Standards*. Washington, DC: National Academy Press, p. 22. www.nap.edu/readingroom/books/nses
4. M.G. Fullan and S. Stiegelbauer. 1991. *The New Meaning of Educational Change*, 2nd ed. New York: Teachers College Press, Columbia University.
   G.E. Hall and S.M. Hord. 1987. *Change in Schools: Facilitating the Process*. Albany: State University of New York Press.
   S. Loucks-Horsley and S. Stiegelbauer. 1991. Using Knowledge of Change to Guide Staff Development. In *Staff Development for Education in the 90s: New Demands, New Realities, New Perspectives*. A. Lieberman and L. Miller, eds. New York: Teachers College Press, Columbia University.
5. See *National Science Education Standards*, p. 158.
6. See *National Science Education Standards*, pp. 55-73.

Worksheet 1: General Overview

1. Terms (fundamental concepts)	Location	Page(s)
evolution		
diversity		
adaptation		
interpreting fossil evidence		
techniques for age determination		
natural selection		
descent from common ancestors		
experiments		
evidence		
explanations		
models		
theory		
skepticism		

Comments on breadth and depth of coverage:

2. Statements of expected student outcomes on evolution and the nature of science	Location	Page(s)
Examples:		
a.		
b.		
c.		

Worksheet 1: (Continued)

3. Student investigations	Location	Page(s)

Titles of example investigations:

a.

b.

c.

Comments:

4. Concept Level	Location	Page(s)

Paragraph 1

Comments:

Paragraph 2

Comments:

Statement of overall impression from the overview:

Worksheet 2: Analysis of Science Subject Matter

A. CONTENT

1. Fundamental understandings addressed: Location Page(s)____

List of fundamental understandings:

2. Do materials promote understanding of the subject matter?

a. Content Standard C: Life Science, or Standard D: Earth and Space Science
Fundamental understanding statement: ______________________ Page(s)____
____

Level of understanding possible based on
the opportunities to learn: Thorough [ ] Some [ ] None [ ]

Comments:

b. Content Standard C: Life Science, or Standard D: Earth and Space Science
Fundamental understanding statement: ______________________ Page(s)____
____

Level of understanding possible based on
the opportunities to learn: Thorough [ ] To some degree [ ] Topic match only [ ]

Comments:

3. Student Investigations

Investigation title: ______________________ Page(s)_____
Learning goal: ______________________
____

The activity alignment between learning goal and *National Science Education Standards* fundamental understanding: Excellent [ ] Partial [ ] None [ ]

Comments:

Worksheet 2: (Continued)

B. SCIENTIFIC INQUIRY

1. What opportunities are provided for students to develop abilities of scientific inquiry?

Cite specific examples: Page(s)____

a. to pose relevant questions; ____

b. plan and conduct investigations; ____

c. use appropriate tools and techniques to gather data; ____

d. use evidence to communicate defensible explanations of cause and effect; ____

e. use scientific criteria to analyze alternative explanations and develop a preferred explanation. ____

Discussion of examples:

2. Opportunities to develop understanding of scientific inquiry: Page(s)____

Cite specific examples:

a. discussion of both roles and limitations of skills such as organizing and interpreting data, constructing explanations; ____

b. discussion of how science advances through legitimate skepticism; ____

c. discussion of how scientists evaluate proposed explanations of others by examining and comparing evidence, reasoning that goes beyond the evidence, suggesting alternative explanations for the same evidence; ____

d. opportunities for students to demonstrate these same understandings as a part of their investigations. ____

Discussion of examples:

Overall estimate of alignment with National Science Education Standards *Inquiry Standard*:
Excellent [ ] Good [ ] Some [ ] Little [ ] None [ ]

Justification of alignment estimate:

Worksheet 2: (Continued)

C. HISTORY AND NATURE OF SCIENCE

Cite specific examples of:

1. evidence supporting the role of scientists, human insight, and scientific reasoning in the historical development of explanations for evolution; Page(s)_____

2. narrative and learning activities that provide examples of how explanations are developed, reviewed by peers, and revised in light of new evidence and thinking; _____

3. specific reference to historical contributions of scientists in the development of fundamental understandings of evolution; _____

4. opportunities for students to demonstrate how scientific explanations are developed, reviewed by peers, and revised in light of new evidence and thinking. _____

Discussion of examples:

Overall estimate of alignment with National Science Education Standards *History and the Nature of Science Standard*
Excellent [ ] Good [ ] Some [ ] Little [ ] None [ ]

Justification of alignment estimate:

Worksheet 3: Analysis of Pedagogy

Cite specific examples where:

1. student learning goals or outcomes focus on one or more fundamental understandings in evolution and the nature of science specified in Content Standards A, C, D, and G; Page(s)_____

   Comments:

2. materials engage and focus student thinking on interesting questions, problems, or relevant issues; rather than opening with statements of fact and vocabulary; _____

   Comments:

3. materials provide a sequence of learning activities connected in such a way as to help students build understanding of a fundamental concept. _____

   Does the material provide specific means (e.g., connections among activities, linkage between text and activities, building from concepts to abstract and embedded assessments) to help the teacher keep students focused on the purpose of the lesson? Yes _______ No______ _____

   Comments:

4. teacher's guide presents common student misconceptions about evolution and the nature of science; _____

   suggestions are provided to access prior understandings of students; and _____

   student learning activities are designed to help students confront misconceptions and encourage conceptual change. _____

   Comments:

Overall estimate of alignment to National Science Education Standards *Teaching Standard*
Excellent [ ] Good [ ] Some [ ] Little [ ] None [ ]

Justification of alignment estimate:

Worksheet 4: Analysis of Assessment Process

Cite example or evidence of:

1. consistency between learning goals and assessment; Page(s)_____

2. assessments stressing application of concepts to new or different situations; _____

3. fairness of assessment tasks for all students—for example, task does not rely too heavily upon the student's ability to read complex items or write explanations, as opposed to understanding the fundamental concepts; and _____

4. the inclusion of actual assessment instruments, scoring criteria or rubrics, and specific suggestions provided regarding their use. _____

Comments:

Overall estimate of alignment to National Science Education Standards *Assessment Standard:*
Excellent [ ] Good [ ] Some [ ] Little or None [ ]

Explanation of alignment estimate:

Worksheet 5: Analysis of Use and Management

1. How many different types of materials must be managed and orchestrated during a typical chapter, unit, or teaching sequence (e.g., student text, teachers guide, transparencies, handouts, videos, software)? Page(s)_____

   Comments:

2. Does the guide contain suggestions for effectively managing instructional materials? _____

3. Do the instructional materials call for equipment, supplies, and technology that teachers using these materials might not have? _____

   Comments:

Overall estimate of use and management:
Easy [ ] Satisfactory [ ] Difficult [ ]

Explanation of overall estimate:

Appendix A

Six Significant Court Decisions Regarding Evolution and Creationism Issues[1]

The following are excerpts from important court decisions regarding evolution and creationism issues. The reader is encouraged to read the full statements as need and time allows.

1. In 1968, in *Epperson v. Arkansas*, the United States Supreme Court invalidated an Arkansas statute that prohibited the teaching of evolution. The Court held the statute unconstitutional on grounds that the First Amendment to the U.S. Constitution does not permit a state to require that teaching and learning must be tailored to the principles or prohibitions of any particular religious sect or doctrine. (*Epperson v. Arkansas*, 393 U.S. 97. (1968))

2. In 1981, in *Segraves v. State of California*, the Court found that the California State Board of Education's *Science Framework*, as written and as qualified by its anti-dogmatism policy, gave sufficient accommodation to the views of Segraves, contrary to his contention that class discussion of evolution prohibited his and his children's free exercise of religion. The anti-dogmatism policy provided that class distinctions of origins should emphasize that scientific explanations focus on "how," not "ultimate cause," and that any speculative statements concerning origins, both in texts and in classes, should be presented conditionally, not dogmatically. The court's ruling also directed the Board of Education to widely disseminate the policy, which in 1989 was expanded to cover all areas of science, not just those concerning issues of origins. (*Segraves v. California*, No. 278978 Sacramento Superior Court (1981))

3. In 1982, in *McLean v. Arkansas Board of Education*, a federal court held that a "balanced treatment" statute violated the Establishment Clause of the U.S. Constitution. The Arkansas statute required public schools to give balanced treatment to "creation-science" and "evolution-science." In a decision that gave a detailed definition of the term "science," the court declared that "creation science" is not in fact a science. The court also found that the statute did not have a secular purpose, noting that the statute used language peculiar to creationist literature in emphasizing origins of life as an aspect of the theory of evolution. While the subject of life's origins is within the province of biology, the scientific community does not consider the subject as part of evolutionary theory, which assumes the existence of life and is directed to an explanation of how life evolved after it originated. The theory of evolution does not presuppose either the absence or the presence of a creator. (*McLean v. Arkansas Board of Education*, 529 F. Supp. 1255, 50 (1982) U.S. Law Week 2412)

4. In 1987, in *Edwards v. Aguillard*, the U.S. Supreme Court held unconstitutional Louisiana's "Creationism Act." This statute prohibited the teaching of evolution in public schools, except when it was accompanied by instruction in "creation science." The Court found that, by advancing the religious belief that a supernatural being created humankind, which is embraced by the term *creation science*, the act impermissibly endorses religion. In addition, the Court found that the provision of a comprehensive science education is undermined when it is forbidden to teach evolution except when creation science is also taught. (*Edwards v. Aguillard*, 482, U.S. 578, 55 (1987) U.S. Law Week 4860, S. CT. 2573, 96 L. Ed. 2d510)

5. In 1990, in *Webster v. New Lennox School District*, the Seventh Circuit Court of Appeals found that a school district may prohibit a teacher from teaching creation science in fulfilling its responsibility to ensure that the First Amendment's establishment clause is not violated, and religious beliefs are not injected into the public school curriculum. The court upheld a

district court finding that the school district had not violated Webster's free speech rights when it prohibited him from teaching "creation science," since it is a form of religious advocacy. (*Webster v. New Lennox School District #122*, 917 F.2d 1004 (7th. Cir., 1990))

6. In 1994, in *Peloza v. Capistrano Unified School District*, the Ninth Circuit Court of Appeals upheld a district court finding that a teacher's First Amendment right to free exercise of religion is not violated by a school district's requirement that evolution be taught in biology classes. Rejecting plaintiff Peloza's definition of a "religion" of "evolutionism," the Court found that the district had simply and appropriately required a science teacher to teach a scientific theory in biology class. (*Peloza v. Capistrano Unified School District*, 37 F.3d 517 (9th Cir., 1994))

NOTE

1. Matsumura, M., ed. 1995. Pp. 2-3 in *Voices for Evolution*. 2nd ed. Berkeley, CA: National Center for Science Education.

Appendix B

Excerpt from "Religion in the Public Schools: A Joint Statement of Current Law"[2]

Schools may teach about explanations of life on earth, including religious ones (such as "creationism"), in comparative religion or social studies classes. In science class, however, they may present only genuinely scientific critiques of, or evidence for, any explanation of life on earth, but not religious critiques (beliefs unverifiable by scientific methodology). Schools may not refuse to teach evolutionary theory in order to avoid giving offense to religion nor may they circumvent these rules by labeling as science an article of religious faith. Public schools must not teach as scientific fact or theory any religious doctrine, including "creationism," although any genuinely scientific evidence for or against any explanation of life may be taught. Just as they may neither advance nor inhibit any religious doctrine, teachers should not ridicule, for example, a student's religious explanation for life on earth.

Note

2. Excerpt from the brochure, "Religion in the Public Schools: A Joint Statement of Current Law." April 1995. Full copy available by contacting Religion in the Public Schools, 15 East 84th Street, Suite 501, New York, NY 10028 or by the World Wide Web at www.ed.gov./Speeches/04-1995/prayer.html. Drafting Committee: American Jewish Congress, Chair; American Civil Liberties Union; American Jewish Committee; American Muslim Council; Anti-Defamation League; Baptist Joint Committee; Christian Legal Society; General Conference of Seventh-Day Adventists; National Association of Evangelicals; National Council of Churches; People for the American Way; Union of American Hebrew Congregations. Endorsing Organizations: American Ethical Union; American Humanist Association; Americans for Religious Liberty; Americans United for Separation of Church and State; B'nai B'rith International; Christian Science Church; Church of the Brethren, Washington Office; Church of Scientology International; Evangelical Lutheran Church in America, Lutheran Office of Governmental Affairs; Federation of Reconstructionist Congregations and Havurot; Friends Committee on National Legislation; Guru Gobind Singh Foundation; Hadassah, The Women's Zionist Organization of America; Interfaith Alliance; Interfaith Impact for Justice and Peace; National Council of Jewish Women; National Jewish Community Relations Advisory Council (NJCRAC); National Ministries, American Baptist Churches, USA; National Sikh Center; North American Council for Muslim Women; Presbyterian Church (USA); Reorganized Church of Jesus Christ of Latter Day Saints; Unitarian Universalist Association of Congregations; United Church of Christ, Office for Church in Society.

Appendix C

Three Statements in Support of Teaching Evolution from Science and Science Education Organizations

1. A NSTA (National Science Teachers Association) Position Statement on the Teaching of Evolution[3]

Approved by the NSTA Board of Directors, July 1997

Introductory Remarks

The National Science Teachers Association supports the position that evolution is a major unifying concept of science and should be included as part of K—College science frameworks and curricula. NSTA recognizes that evolution has not been emphasized in science curricula in a manner commensurate to its importance because of official policies, intimidation of science teachers, the general public's misunderstanding of evolutionary theory, and a century of controversy.

Furthermore, teachers are being pressured to introduce creationism, creation "science," and other nonscientific views, which are intended to weaken or eliminate the teaching of evolution.

Within this context, NSTA recommends that:

- Science curricula and teachers should emphasize evolution in a manner commensurate with its importance as a unifying concept in science and its overall explanatory power.

- Policy-makers and administrators should not mandate policies requiring the teaching of creation science or related concepts such as so-called "intelligent design," "abrupt appearance," and "arguments against evolution."

- Science teachers should not advocate any religious view about creation, nor advocate the converse: that there is no possibility of supernatural influence in bringing about the universe as we know it. Teachers should be nonjudgmental about the personal beliefs of students.

- Administrators should provide support to teachers as they design and implement curricula that emphasize evolution. This should include inservice education to assist teachers to teach evolution in a comprehensive and professional manner. Administrators also should support teachers against pressure to promote nonscientific views or to diminish or eliminate the study of evolution.

- Parental and community involvement in establishing the goals of science education and the curriculum development process should be encouraged and nurtured in our democratic society. However, the professional responsibility of science teachers and curriculum specialists to provide students with quality science education should not be bound by censorship, pseudoscience, inconsistencies, faulty scholarship, or unconstitutional mandates.

- Science text books shall emphasize evolution as a unifying concept. Publishers should not be required or volunteer to include disclaimers in textbooks concerning the nature and study of evolution.

NSTA offers the following background information:

The Nature of Science and Scientific Theories

Science is a method of explaining the natural world. It assumes the universe operates according to regularities and that through systematic investigation we can understand these regularities. The methodology of science emphasizes the logical testing of alternate explanations of natural phenomena against empirical data. Because science is limited to explaining the natural world by means of natural processes, it cannot use supernatural causation in its explanations. Similarly, science is precluded from making statements about supernatural forces because these are outside its provenance. Science has increased our knowledge because of this insistence on the search for natural causes.

The most important scientific explanations are called "theories." In ordinary speech, "theory" is often used to mean "guess," or "hunch," whereas in scientific terminology, a theory is a set of universal statements which explain the natural world. Theories are powerful tools. Scientists seek to develop theories that

- are internally consistent and compatible with the evidence
- are firmly grounded in and based upon evidence
- have been tested against a diverse range of phenomena
- possess broad and demonstrable effectiveness in problem solving
- explain a wide variety of phenomena.

The body of scientific knowledge changes as new observations and discoveries are made. Theories and other explanations change. New theories emerge and other theories are modified or discarded. Throughout this process, theories are formulated and tested on the basis of evidence, internal consistency, and their explanatory power.

Evolution as a Unifying Concept

Evolution in the broadest sense can be defined as the idea that the universe has a history: that change through time has taken place. If we look today at the galaxies, stars, the planet earth, and the life on planet earth, we see that things today are different from what they were in the past: galaxies, stars, planets, and life forms have evolved. Biological evolution refers to the scientific theory that living things share ancestors from which they have diverged: Darwin called it "descent with modification." There is abundant and consistent evidence from astronomy, physics, biochemistry, geochronology, geology, biology, anthropology, and other sciences that evolution has taken place.

As such, evolution is a unifying concept for science. The *National Science Education Standards* recognizes that conceptual schemes such as evolution "unify science disciplines and provide students with powerful ideas to help them understand the natural world," and recommends evolution as one such scheme. In addition, the *Benchmarks for Science Literacy* from the American Association for the Advancement of Science's Project 2061 and NSTA's Scope, Sequence, and Coordination Project, as well as other national calls for science reform, all name evolution as a unifying concept because of its importance across the discipline of science. Scientific disciplines with a historical component, such as astronomy, geology, biology, and anthropology, cannot be taught with integrity if evolution is not emphasized.

There is no longer a debate among scientists over whether evolution has taken place. There is considerable debate about how evolution has taken place: the processes and mechanisms producing change, and what has happened during the history of the universe. Scientists often disagree about their explanations. In any science, disagreements are subject to rules of evaluation. Errors and false conclusions are confronted by experiment and observation, and evolution, as in any aspect of science, is continually open to and subject to experimentation and questioning.

Creationism

The word "creationism" has many meanings. In its broadest meaning, creationism is the idea that a supernatural power or powers created. Thus to Christians, Jews, and Muslims, God created; to the Navajo, the Hero Twins created. In a narrower sense, "creationism" has come to mean "special creation": the doctrine that the universe and all that is in it was created by God in essentially its present form, at one time. The most common variety of special creationism asserts that

- the earth is very young
- life was originated by a creator
- life appeared suddenly
- kinds of organisms have not changed
- all life was designed for certain functions and purposes.

This version of special creation is derived from a literal interpretation of Biblical Genesis. It is a specific, sectarian religious belief that is not held by all religious people. Many Christians and Jews believe that God created through the process of evolution. Pope John Paul II, for example, issued a statement in 1996 that reiterated the Catholic position that God created, but that the scientific evidence for evolution is strong.

"Creation science" is an effort to support special creationism through methods of science. Teachers are often pressured to include it or synonyms such as "intelligent design theory," "abrupt appearance theory," "initial complexity theory," or "arguments against evolution" when they teach evolution. Special creationist claims have been discredited by the available evidence. They have no power to explain the natural world and its diverse phenomena. Instead, creationists seek out supposed anomalies among many existing theories and accepted facts. Furthermore, creation science claims do not provide a basis for solving old or new problems or for acquiring new information.

Nevertheless, as noted in the *National Science Education Standards*, "Explanations on how the natural world changed based on myths, personal beliefs, religious values, mystical inspiration, superstition, or authority may be personally useful and socially relevant, but they are not scientific." Because science can only use natural explanations and not supernatural ones, science teachers should not advocate any religious view about creation, nor advocate the converse: that there is no possibility of supernatural influence in bringing about the universe as we know it.

Legal Issues

Several judicial rulings have clarified issues surrounding the teaching of evolution and the imposition of mandates that creation science be taught when evolution is taught. The First Amendment of the Constitution requires that public institutions such as schools be religiously neutral; because special creation is a specific, sectarian religious view, it cannot be advocated as "true," accurate scholarship in the public schools. When Arkansas passed a law requiring "equal time" for creationism and evolution, the law was challenged in Federal District Court. Opponents of the bill included the religious leaders of the United Methodist, Episcopalian, Roman Catholic, African Methodist Episcopal, Presbyterian, and Southern Baptist churches, and several educational organizations. After a full trial, the judge ruled that creation science did not qualify as a scientific theory (*McLean v. Arkansas Board of Education*, 529 F. Supp. 1255 (ED Ark. 1982)).

Louisiana's equal time law was challenged in court and eventually reached the Supreme Court. In *Edwards v. Aguillard* 482 U.S. 578 (1987), the court determined that creationism was inherently a religious idea and to mandate or advocate it in the public schools would be unconstitutional. Other court decisions have upheld the right of a district to require that a teacher teach evolution and not teach creation science: (*Webster v. New Lennox School District #122*, 917 F.2d 1003 (7th Cir. 1990); *Peloza v. Capistrano Unified School District*, 37 F.3d 517 (9th Cir. 1994)).

Some legislatures and policy-makers continue attempts to distort the teaching of evolution through mandates that would require teachers to teach evolution as "only a theory," or that require a textbook or lesson on evolution to be preceded by a disclaimer. Regardless of the legal status of these mandates, they are bad educational policy. Such policies have the effect of intimidating teachers, which may result in the de-emphasis or omission of evolution. The public will only be further confused about the special nature of scientific theories, and if less evolution is learned by students, science literacy itself will suffer.

References

American Association for the Advancement of Science (AAAS). 1993. *Benchmarks for Science Literacy*. Project 2061. New York: Oxford University Press.

Daniel v. Waters. 515 F.2d 485 (6th Cir., 1975).

Edwards v. Aguillard. 482 U.S. 578 (1987).

Epperson v. Arkansas. 393 U.S. 97 (1968)

Laudan, Larry. 1996. *Beyond Positivism and Relativism: Theory, Method, and Evidence*. Boulder, CO: Westview Press.

McLean v. Arkansas Board of Education. 529 F. Supp. 1255 (D. Ark. 1982).

National Research Council (NRC). 1996. *National Science Education Standards*. Washington, DC: National Academy Press.

National Science Teachers Association (NSTA). 1996. *A Framework for High School Science Education*. Arlington, VA: National Science Teachers Association.

NSTA. 1993. *The Content Core: Vol. I.* Rev. ed. Arlington, VA: National Science Teachers Association.

Peloza v. Capistrano Unified School District. 37 F.3d 517 (9th Cir. 1994).

Ruse, Michael. 1996. *But Is It Science? The Philosophical Question in the Creation/Evolution Controversy*. Amherst, NY: Prometheus Books.

Webster v. New Lennox School District #122. 917 F.2d 1003 (7th Cir. 1990).

Task Force Members

Gerald Skoog, Chair, College of Education, Texas Tech University, Lubbock, Texas

Randy Cielen, Joseph Teres School, Winnipeg, Manitoba, Canada

Linda Jordan, Science Consultant, Franklin, Tennessee

Janis Lariviere, Westlake Alternative Learning Center, Austin, Texas

Larry Scharmann, Kansas State University, Manhattan, Kansas

Eugenie Scott, National Center for Science Education, Berkeley, California

2. National Association of Biology Teachers Statement on Teaching Evolution[4]

As stated in *The American Biology Teacher* by the eminent scientist Theodosius Dobzhansky (1973), "Nothing in biology makes sense except in the light of evolution."[5] This often-quoted assertion accurately illuminates the central, unifying role of evolution in nature, and therefore in biology. Teaching biology in an effective and scientifically-honest manner requires classroom discussions and laboratory experiences on evolution.

Modern biologists constantly study, ponder and deliberate the patterns, mechanisms and pace of evolution, but they do not debate evolution's occurrence. The fossil record and the diversity of extant organisms, combined with modern techniques of molecular biology, taxonomy and geology, provide exhaustive examples and powerful evidence for genetic variation, natural selection, speciation, extinction and other well-established components of current evolutionary theory. Scientific deliberations and modifications of these components clearly demonstrate the vitality and scientific integrity of evolution and the theory that explains it.

The same examination, pondering and possible revision have firmly established evolution as an important natural process explained by valid scientific principles, and clearly differentiate and separate science from various kinds of nonscientific ways of knowing, including those with a supernatural basis such as creationism. Whether called "creation science," "scientific creationism," "intelligent-design theory," "young-earth theory" or some other synonym, creation beliefs have no place in the science classroom. Explanations employing nonnaturalistic or supernatural events, whether or not explicit reference is made to a supernatural being, are outside the realm of science and not part of a valid science curriculum. Evolutionary theory, indeed all of science, is necessarily silent on religion and neither refutes nor supports the existence of a deity or deities.

Accordingly, the National Association of Biology Teachers, an organization of science teachers, endorses the following tenets of science, evolution and biology education:

- The diversity of life on earth is the outcome of evolution: an unpredictable and natural process of temporal descent with genetic modification that is affected by natural selection, chance, historical contingencies and changing environments.
- Evolutionary theory is significant in biology, among other reasons, for its unifying properties and predictive features, the clear empirical testability of its integral models, and the richness of new scientific research it fosters.
- The fossil record, which includes abundant transitional forms in diverse taxonomic groups, establishes extensive and comprehensive evidence for organic evolution.
- Natural selection, the primary mechanism for evolutionary changes, can be demonstrated with numerous, convincing examples, both extant and extinct.
- Natural selection—a differential, greater survival and reproduction of some genetic variants within a population under an existing environmental state—has no specific direction or goal, including survival of a species.
- Adaptations do not always provide an obvious selective advantage. Furthermore, there is no indication that adaptations—molecular to organismal—must be perfect: adaptations providing a selective advantage must simply be good enough for survival and increased reproductive fitness.
- The model of punctuated equilibrium provides another account of the tempo of speciation in the fossil record of many lineages: it does not refute or overturn evolutionary theory, but instead adds to its scientific richness.
- Evolution does not violate the second law of thermodynamics: producing order from disorder is possible with the addition of energy, such as from the sun.
- Although comprehending deep time is difficult, the earth is about 4.5 billion years old. *Homo sapiens* has occupied only a minuscule moment of that immense duration of time.
- When compared with earlier periods, the Cambrian explosion evident in the fossil record reflects at least three phenomena: the evolution of animals with readily fossilized hard body parts; Cambrian environment (sedimentary rock) more conducive to preserving fossils; and the evolution from pre-Cambrian forms of an increased diversity of body patterns in animals.
- Radiometric and other dating techniques, when used properly, are highly accurate means of establishing dates in the history of the planet and in the history of life.
- In science, a theory is not a guess or an approximation but an extensive explanation developed from well-documented, reproducible sets of experimentally-derived data from repeated observations of natural processes.
- The models and the subsequent outcomes of a scientific theory are not decided in advance, but can be, and often are, modified and improved as new empirical evidence is uncovered. Thus, science is a constantly self-correcting endeavor to understand nature and natural phenomena.

- Science is not teleological: the accepted processes do not start with a conclusion, then refuse to change it, or acknowledge as valid only those data that support an unyielding conclusion. Science does not base theories on an untestable collection of dogmatic proposals. Instead, the processes of science are characterized by asking questions, proposing hypotheses, and designing empirical models and conceptual frameworks for research about natural events.
- Providing a rational, coherent and scientific account of the taxonomic history and diversity of organisms requires inclusion of the mechanisms and principles of evolution.
- Similarly, effective teaching of cellular and molecular biology requires inclusion of evolution.
- Specific textbook chapters on evolution should be included in biology curricula, and evolution should be a recurrent theme throughout biology textbooks and courses.
- Students can maintain their religious beliefs and learn the scientific foundations of evolution.
- Teachers should respect diverse beliefs, but contrasting science with religion, such as belief in creationism, is not a role of science. Science teachers can, and often do, hold devout religious beliefs, accept evolution as a valid scientific theory, and teach the theory's mechanisms and principles.
- Science and religion differ in significant ways that make it inappropriate to teach any of the different religious beliefs in the science classroom.

Opposition to teaching evolution reflects confusion about the nature and processes of science. Teachers can, and should, stand firm and teach good science with the acknowledged support of the courts. In *Epperson v. Arkansas* (1968), the U.S. Supreme Court struck down a 1928 Arkansas law prohibiting the teaching of evolution in state schools. In *McLean v. Arkansas* (1982), the federal district court invalidated a state statute requiring equal classroom time for evolution and creationism.

Edwards v. Aguillard (1987) led to another Supreme Court ruling against so-called "balanced treatment" of creation science and evolution in public schools. In this landmark case, the Court called the Louisiana equal-time statute "facially invalid as violative of the Establishment Clause of the First Amendment, because it lacks a clear secular purpose." This decision—"the Edwards restriction"—is now the controlling legal position on attempts to mandate the teaching of creationism: the nation's highest court has said that such mandates are unconstitutional. Subsequent district court decisions in Illinois and California have applied "the Edwards restriction" to teachers who advocate creation science, and to the right of a district to prohibit an individual teacher from promoting creation science, in the classroom.

Courts have thus restricted school districts from requiring creation science in the science curriculum and have restricted individual instructors from teaching it. All teachers and administrators should be mindful of these court cases, remembering that the law, science and NABT support them as they appropriately include the teaching of evolution in the science curriculum.

References and Suggested Reading

Clough, M. 1994. Diminish students' resistance to biological evolution. *American Biology Teacher* 56(Oct.):409-415.

Futuyma, D. 1997. *Evolutionary Biology*. 3rd ed. Sunderland, MA: Sinauer Associates, Inc.

Gillis, A. 1994. Keeping creationism out of the classroom. *BioScience* 44:650-656.

Gould, S. 1994. The evolution of life on the earth. *Scientific American* 271(Oct.):85-91.

Gould, S. 1977. *Ever Since Darwin: Reflections in Natural History*. New York: W.W. Norton.

Mayr, E. 1991. *One Long Argument: Charles Darwin and the Genesis of Modern Evolutionary Thought.* Cambridge, MA: Harvard University Press.

McComas, W., ed. 1994. *Investigating Evolutionary Biology in the Laboratory.* Reston, VA: National Association of Biology Teachers.

Moore, J. 1993. *Science as a Way of Knowing: The Foundation of Modern Biology.* Cambridge, MA: Harvard University Press.

National Center for Science Education, P.O. Box 9477, Berkeley, CA 94709. Numerous publications such as *Facts, faith and fairness: Scientific creationism clouds scientific literacy* by S. Walsh and T. Demere.

Numbers, R. 1993. *The Creationists: The Evolution of Scientific Creationism.* Berkeley, CA: University of California Press.

Weiner, J. 1994. The *Beak of the Finch: A Story of Evolution in Our Time.* New York: Alfred A. Knopf.

3. Resolution passed by the American Association for the Advancement of Science Commission on Science Education[6]

The Commission on Science Education of the American Association for the Advancement of Science, is vigorously opposed to attempts by some boards of education, and other groups, to require that religious accounts of creation be taught in science classes.

During the past century and a half, the earth's crust and the fossils preserved in it have been intensively studied by geologists and paleontologists. Biologists have intensively studied the origin, structure, physiology, and genetics of living organisms. The conclusion of these studies is that the living species of animals and plants have evolved from different species that lived in the past. The scientists involved in these studies have built up the body of knowledge known as the biological theory of the origin and evolution of life. There is no currently acceptable alternative scientific theory to explain the phenomena.

The various accounts of creation that are part of the religious heritage of many people are not scientific statements or theories. They are statements that one may choose to believe, but if he does, this is a matter of faith, because such statements are not subject to study or verification by the procedures of science. A scientific statement must be capable of test by observation and experiment. It is acceptable only if, after repeated testing, it is found to account satisfactorily for the phenomena to which it is applied.

Thus the statements about creation that are part of many religions have no place in the domain of science and should not be regarded as reasonable alternatives to scientific explanations for the origin and evolution of life.

Resolution on Inclusion of the Theory of Creation in Science Curricula[7]

WHEREAS some State Boards of Education and State Legislatures have required or are considering requiring inclusion of the theory of creation as an alternative to evolutionary theory in discussions of origins of life, and

WHEREAS the requirement that the theory of creation be included in textbooks as an alternative to evolutionary theory represents a constraint upon the freedom of the science teacher in the classroom, and

WHEREAS its inclusion also represents dictation by a lay body of what shall be considered within the corpus of a science,

THEREFORE the American Association for the Advancement of Science strongly urges that reference to the theory of creation, which is neither scientifically grounded nor capable of performing the roles required of scientific theories, not be required in textbooks and other classroom materials intended for use in science curricula.

Statement on Forced Teaching of Creationist Beliefs in Public School Science Education[8]

WHEREAS it is the responsibility of the American Association for the Advancement of Science to preserve the integrity of science, and

WHEREAS science is a systematic method of investigation based on continuous experimentation, observation, and measurement leading to evolving explanations of natural phenomena, explanations which are continuously open to further testing, and

WHEREAS evolution fully satisfies these criteria, irrespective of remaining debates concerning its detailed mechanisms, and

WHEREAS the Association respects the right of people to hold diverse beliefs about creation that do not come within the definitions of science, and

WHEREAS creationist groups are imposing beliefs disguised as science upon teachers and students to the detriment and distortion of public education in the United States,

THEREFORE be it resolved that because "creationist science" has no scientific validity it should not be taught as science, and further, that the AAAS views legislation requiring "creationist science" to be taught in public schools as a real and present threat to the integrity of education and the teaching of science, and

Be it further resolved that the AAAS urges citizens, educational authorities, and legislators to oppose the compulsory inclusion in science education curricula of beliefs that are not amenable to the process of scrutiny, testing, and revision that is indispensable to science.

NOTES

3. Reprinted with permission from NSTA Publications, copyright 1997 from NSTA Handbook, 1997-98, National Science Teachers Association, 1840 Wilson Boulevard, Arlington, VA 22201-3000.
4. Statement on Teaching Evolution, National Association of Biology Teachers (NABT). Adopted by the NABT Board of Directors on March 15, 1995.
5. Dobzhansky, T. 1973. Nothing in biology makes sense except in the light of evolution. *American Biology Teacher* 35:125-129.
6. American Association for the Advancement of Science (AAAS), Commission on Science Education. October 13, 1972.
7. Adopted by AAAS Council on December 30, 1972.
8. Adopted by the AAAS Board of Directors on January 4, 1982, and by the AAAS Council on January 7, 1982.

Appendix D

References for Further Reading and Other Resources

The following list of references represents a sampling of the vast literature available on education, biology, and evolution. The reader is encouraged to explore the literature further as need and time allow.

Please visit our World Wide Web address at http://www4.nas.edu/opus/evolve.nsf for more extensive resource listings for these subjects.

Publications on Education

AAAS (American Association for the Advancement of Science). 1993. *Benchmarks for Science Literacy*. Project 2061. New York: Oxford University Press.

Bybee, R. 1997. *Achieving Scientific Literacy: From Purposes to Practices*. Portsmouth, NH: Heinemann Educational Books.

Bybee, R. 1996. *National Standards and the Science Curriculum: Challenges, Opportunities, and Recommendations.* Dubuque, IA: Kendall/Hunt Publishing Co.

NRC (National Research Council). 1996. *National Science Education Standards*. Washington, DC: National Academy Press.

NSRC (National Science Resources Center). 1997. *Science for All Children: A Guide to Improving Elementary Science Education in Your School District*. Washington, DC: National Academy Press.

NSTA (National Science Teachers Association). 1996. *A Framework for High School Science Education*. Arlington, VA: National Science Teachers Association.

NSTA. 1993. *Scope, Sequence, and Coordination of Secondary School Science. Vol. I. The Content Core: A Guide for Curriculum Designers*. rev. ed. Arlington, VA: National Science Teachers Association.

Publications on Biology and Other Sciences

Berg, P., and M. Singer. 1992. *Dealing with Genes: The Language of Heredity*. Mill Valley, CA: University Science Books.

BSCS (Biological Sciences Curriculum Study). 1998. *BSCS Biology: An Ecological Approach*. 8th ed. Dubuque, IA: Kendall/Hunt Publishing Co.

BSCS. 1997. *BSCS Biology: A Human Approach*. Dubuque, IA: Kendall/Hunt Publishing Co.

BSCS. 1996. *Biological Science: A Molecular Approach*. 7th ed. Lexington, MA: D.C. Heath.

BSCS. 1993. *Developing Biological Literacy: A Guide to Developing Secondary and Post-secondary Biology Curricula*. Colorado Springs, CO: BSCS.

BSCS. 1983. *Biological Science: Interaction of Experiments and Ideas*. Englewood Cliffs, NJ: Prentice Hall.

BSCS. 1978. *Biology Teachers' Handbook*. 3rd ed. William V. Mayer, ed. New York: John Wiley and Sons.

Campbell, N. 1996. *Biology*. 4th ed. Menlo Park, CA: Benjamin-Cummings.

ESCP (Earth Science Curriculum Project). 1973. *Investigating the Earth*. rev. ed. Boston, MA: Houghton Mifflin.

Jacob, F. 1982. *The Possible and the Actual*. New York: Pantheon Books.

Mayr, E. 1997. *This Is Biology: The Science of the Living World.* Cambridge, MA: Belknap Press of Harvard University Press.

Moore, J.A. 1993. *Science as a Way of Knowing: The Foundations of Modern Biology*. Cambridge, MA: Harvard University Press.

Oosterman, M., and M. Schmidt, eds. 1990. *Earth Science Investigations*. Alexandria, VA: American Geological Institute.

Raven, P.H., and G.B. Johnson. 1992. *Biology*. 3rd ed. St. Louis, MO: Mosby Year Book, Inc.

Scientific American. 1994. Life in the universe: special issue. 271(Oct.).

Trefil, J., and R.M. Hazen. 1998. *The Sciences: An Integrated Approach*. 2nd ed. New York: John Wiley and Sons.

Publications on Evolution

Berra, T. 1990. *Evolution and the Myth of Creationism: A Basic Guide to the Facts in the Evolution Debate*. Stanford, CA: Stanford University Press.

Clough, M. 1994. Diminish students' resistance to biological evolution. *American Biology Teacher* 56:409–415.

Darwin, C. 1934. *Charles Darwin's Diary of the Voyage of H.M.S. Beagle*, Nora Barlow, ed. Cambridge, UK: The University Press.

Darwin, C. 1859. *On the Origin of Species by Means of Natural Selection*. London: J. Murray.

Dawkins, R. 1996. *Climbing Mount Improbable*. New York: W.W. Norton.

Dawkins, R. 1986. *The Blind Watchmaker: Why Evidence of Evolution Reveals a Universe Without Design*. New York: W.W. Norton.

de Duve, C. 1995. *Vital Dust: Life as a Cosmic Imperative*. New York: Basic Books.

Dennett, D.C. 1995. *Darwin's Dangerous Idea: Evolution and the Meanings of Life*. New York: Simon and Schuster.

Diamond, J. 1997. *Guns, Germs, and Steel: The Fates of Human Societies*. New York: W.W. Norton.

Diamond, J. 1992. *The Third Chimpanzee: The Evolution and Future of the Human Animal*. New York: HarperCollins.

Diamond, J., and M.L. Cody, eds. 1975. *Ecology and Evolution of Communities*. Cambridge, MA: Belknap Press of Harvard University Press.

Ewald, P. 1994. *The Evolution of Infectious Disease*. New York: Oxford University Press.

Futuyma, D. 1997. *Evolutionary Biology*. 3rd ed. Sunderland, MA: Sinauer Associates, Inc.

Futuyma, D. 1995. *Science on Trial: The Case for Evolution*. 2nd ed. Sunderland, MA.: Sinauer Associates, Inc.

Gillis, A. 1994. Keeping creationism out of the classroom. *BioScience* 44:650-656.

Goldschmidt, T. 1996. *Darwin's Dreampond: Drama in Lake Victoria*. Cambridge, MA: MIT Press.

Goldsmith, T. H. 1991. *The Biological Roots of Human Nature: Forging Links Between Evolution and Behavior*. New York: Oxford University Press.

Gould, S.J. 1997. This view of life: Nonoverlapping magisteria. *Natural History* 106(2):16-22.

Gould, S.J. 1994. The evolution of life on the earth. *Scientific American* 271(Oct):85-91.

Gould, S.J. 1989. *Wonderful Life: The Burgess Shale and the Nature of History*. New York: W.W. Norton.

Gould, S.J. 1980. *The Panda's Thumb: More Reflections in Natural History*. New York: W.W. Norton.

Gould, S.J. 1977. *Ever Since Darwin: Reflections in Natural History*. New York: W.W. Norton.

Kitcher, P. 1982. *Abusing Science: The Case Against Creationism*. Cambridge, MA: MIT Press.

Matsumura, M., ed. 1995. *Voices for Evolution*. 2nd ed. Berkeley, CA: National Center for Science Education.

Mayr, E. 1991. *One Long Argument: Charles Darwin and the Genesis of Modern Evolutionary Thought*. Cambridge, MA: Harvard University Press.

Mayr, E. 1972. The nature of the Darwinian revolution. *Science* 176:981-989.

McComas, W., ed. 1994. *Investigating Evolutionary Biology in the Laboratory*. Reston, VA: National Association of Biology Teachers.

McKinney, M.L. 1993. *Evolution of Life: Processes, Patterns, and Prospects*. Englewood Cliffs, NJ: Prentice Hall.

Moore, J.R. 1979. *The Post-Darwinian Controversies: A Study of the Protestant Struggle to Come to Terms with Darwin in Great Britain and America, 1870-1900.* Cambridge, UK: Cambridge University Press.

Nesse, R., and G. Williams. 1995. *Why We Get Sick: The New Science of Darwinian Medicine*. New York: Times Books.

Newell, N.D. 1982. *Creation and Evolution: Myth or Reality?* New York: Columbia University Press.

Numbers, R. 1993. *The Creationists: The Evolution of Scientific Creationism*. Berkeley, CA: University of California Press.

Quammen, D. 1996. *The Song of the Dodo: Island Biogeography in an Age of Extinctions*. New York: Scribner.

Ruse, M. 1996. *But Is It Science? The Philosophical Question in the Creation/Evolution Controversy*. Amherst, NY: Prometheus Books.

Ruse, M. 1982. *Darwinism Defended: A Guide to the Evolution Controversies*. Reading, MA: Addison-Wesley.

Ruse, M. 1979. *The Darwinian Revolution: Science Red in Tooth and Claw*. Chicago: University of Chicago Press.

Tiffin, L. 1994. *Creationism's Upside-down Pyramid: How Science Refutes Fundamentalism*. Amherst, NY: Prometheus Books.

Walsh, S., and T. Demere. 1993. *Facts, Faith and Fairness: Scientific Creationism Clouds Scientific Literacy*. Berkeley, CA: National Center for Science Education.

Weiner, J. 1994. *The Beak of the Finch: A Story of Evolution in Our Time.* New York: Alfred A. Knopf.

Wills, C. 1989. *The Wisdom of the Genes: New Pathways in Evolution*. New York: Basic Books.

Wilson, E. 1992. *The Diversity of Life*. Cambridge, MA: Harvard University Press.

Publications on the Nature of Science

Aicken, F. 1991. *The Nature of Science*. 2nd ed. Portsmouth, NH: Heinemann Educational Books.

Bronowski, J. 1965. *Science and Human Values*. New York: Harper.

Chalmers, A. 1995. *What Is This Thing Called Science?* 2nd ed. Indianapolis: Nackett.

Chalmers, A. 1990. *Science and Its Fabrication*. Minneapolis, MN: University of Minnesota Press.

Daedalus. 1978. Limits of scientific inquiry. 107 (Spring).

Hull, D. 1988. *Science as a Process: An Evolutionary Account of the Social and Conceptual Development of Science*. Chicago: University of Chicago Press.

Kuhn, T.S. 1970. *The Structure of Scientific Revolutions*. Chicago: University of Chicago Press.

Laudan, Larry. 1996. *Beyond Positivism and Relativism: Theory, Method, and Evidence*. Boulder, CO: Westview Press.

Popper, K. 1994. *The Myth of the Framework: In Defense of Science and Rationality*. London: Routledge.

Wolpert, L. 1992. *The Unnatural Nature of Science*. Cambridge, MA: Harvard University Press.

Woolgar, S. 1988. *Science: The Very Idea*. London: Routledge.

Videos

The Day the Universe Changed (episode #10, Worlds Without End). 1986. Owings Mills, MD: MPT-TV.

The Pleasure of Finding Things Out. 1982. Video interview with Richard Feynman. New York: Time/Life video.

Darwin's Revolution in Thought. Talk given by Stephen Jay Gould (No. 126). Available from Into the Classroom Video, 351 Pleasant Street, Northhampton, MA 01060.

God, Darwin and the Dinosaurs. 1989. Boston: WGBH Educational Foundation.

In the Beginning: The Creationist Controversy. 1994. Chicago: WTTW.

Appendix E

Reviewers

This report has been reviewed by individuals chosen for their diverse perspectives and technical expertise, in accordance with procedures approved by the NRC's Report Review Committee. The purpose of this independent review is to provide candid and critical comments that will assist the authors and the NRC in making their published report as sound as possible and to ensure that the report meets institutional standards for objectivity, evidence, and responsiveness to the study charge. The content of the review comments and draft manuscript remain confidential to protect the integrity of the deliberative process. We wish to thank the following individuals for their participation in the review of this report:

Paul Baker
Evan Pugh Professor of Anthropology, Emeritus
Pennsylvania State University
Kaneohe, Hawaii

Howard Berg
Professor of Biology
Harvard University
Cambridge, Massachusetts

Donald Brown
Department of Embryology
Carnegie Institution of Washington
Washington, DC

Wayne Carley
Executive Director
National Association of Biology Teachers
Reston, Virginia

Betty Carvellas
Biology Teacher
Essex High School
Essex Junction, Vermont

Wilford Gardner
Adjunct Professor of Soil Physics
University of California at Berkeley
Berkeley, California

Robert Griffiths
Professor of Physics
Carnegie Mellon University
Pittsburgh, Pennsylvania

Dudley Herschbach
Professor of Science
Harvard University
Cambridge, Massachusetts

Ken Miller
Professor of Biology
Brown University
Providence, Rhode Island

Nancy Ridenour
Biology Teacher and
Science Department Chair
Ithaca High School
Ithaca, New York

Martin Rodbell
Scientist Emeritus
National Institute of Environmental
Health Sciences
Research Triangle Park, North Carolina

Robert Sinsheimer
Professor of Biology, Emeritus
University of California at Santa Barbara
Santa Barbara, California

Gerald Skoog
Helen DeVitt Jones Professor of Curriculum
and Instruction
Texas Technology University
Lubbock, Texas

George Wertherill
Department of Terrestrial Magnetism
Carnegie Institution of Washington
Washington, DC

And other anonymous reviewers.

While the individuals listed above have provided many constructive comments and suggestions, responsibility for the final content of this report rests solely with the authoring committee and the NRC.

Index

V

W

Credits

Cover, title page, pages x and 1. Earth view from space, NASA.

Cover, title page, and page 35: Grand Canyon, Photodisk.

Cover, title page, and page vi: Fossil fish, Photodisk.

Cover, title page, and page 3: Coral reef, Stephen Fink/Corbis.

Cover and title page: Leonardo da Vinci.

Cover and page 55: Nautilus fossil, NAP Image Archives.

page iv: Entrance to National Academy of Sciences Building, Carol M. Highsmith, photographer.

page v: Marble seal of the National Academy of Sciences, David Patterson, photographer.

page 2: Rain forest, Stephen Dalton, photographer, © Oxford Scientific Films.

page 3: Insert, left: Bass, Dallas Aquarium, © The National Audubon Society Collection.

page 3: Insert, right: Fossil fish (*Priscacara oxyprion*), © E. R. Degginger.

page 4: Graphic by Leigh Coriale Design and Illustration, adapted from *Essential Cell Biology*, Garland Publishing, Inc.

page 6: Balcones Research Center, University of Texas at Austin, © 1997 by Phyllis Janik.

page 8: *Archaeopteryx* cast, Smithsonian Museum, Washington, DC, James Amos/Corbis.

page 10: Galapagos Islands view from space, NASA.

pages 11 and 19: Galapagos finch, Galen Rowell/Corbis.

pages 12 and 27: Young stars, Hubble Space Telescope, NASA.

page 13: Charles Darwin, Corbis-Bettmann.

page 13: Alfred Russel Wallace, Library of Congress.

page 13: Gregor Mendel, Corbis-Bettmann.

page 15: Ant in amber, David Grimaldi, American Museum of Natural History, New York, New York.

page 17: Lacewing graphic by Leigh Coriale Design and Illustration.

page 17: Lacewing photograph, Catherine and Maurice Tauber, Cornell University, Ithaca, New York.

page 18: Whale ancestors, drawings by N. Haver, © Sinauer Associates, Inc.

page 20: Skulls, drawings by Darwen Hennings, © Wadsworth Publishing Company.

page 21: Wasp and caterpillar, James H. Tumlinson, U.S. Department of Agriculture.

page 23: Graphic by Leigh Coriale Design and Illustration.

page 26: Armillary sphere, Library of Congress.

page 28: Nicolaus Copernicus, Corbis-Bettmann.

page 28: Johannes Kepler, Library of Congress.

page 28: Galileo Galilei, Library of Congress/Corbis.

page 28: Isaac Newton, Library of Congress.

page 28: 18th century view of the universe, Corbis-Bettmann.

page 30: Tropical forest with crane, Smithsonian Institution.

page 31: Leafnosed bat, Joe McDonald/Corbis.

page 32: Graphic by Leigh Coriale Design and Illustration.

page 33: Sedimentary rocks, David McConnell, University of Akron, Akron, Ohio.

page 34: Fossil record, graphic by Leigh Coriale Design and Illustration, derived from an illustration developed by Ken Miller, Brown University, Providence, Rhode Island.

page 36: Armadillo, Joe McDonald/Corbis.

page 36: Fossil, courtesy of Raymond T. Rye, Smithsonian Institution.

page 36: Stromatolites, courtesy of Embassy of Australia.

page 36-37: Timeline graphic by Leigh Coriale Design and Illustration, adapted from *The Book of Life*, W.W. Norton, New York, New York.

page 38: Graphic by Leigh Coriale Design and Illustration, adapted from *Essential Cell Biology*, Garland Publishing, Inc.

page 39: Graphic by Leigh Coriale Design and Illustration, adapted from *From So Simple a Beginning*, Macmillan Publishing Company, New York, New York.

page 41: Plate tectonics, graphic by Leigh Coriale Design and Illustration, adapted from *Astronomy Today*, Prentice Hall, Englewood Cliffs, New Jersey.

page 44: Graphic by Leigh Coriale Design and Illustration.

page 46: Student at Montgomery Blair High School, Silver Spring, Maryland, Robert Allen Strawn, photographer.
page 49: Students at Piney Branch Elementary School, Takoma Park, Maryland, Robert Allen Strawn, photographer.
page 51: Students at Eastern Middle School, Silver Spring, Maryland, Robert Allen Strawn, photographer.
page 54: Trilobite, David McGrath, photographer, © Photo Archives, Denver Museum of Natural History.
page 60: Students and teacher, Milton Hershey School, Hershey, PA, © Blair Seitz.
page 61: Student at Piney Branch Elementary School, Takoma Park, Maryland, Robert Allen Strawn, photographer.
page 64: Draft Growth-of-Understanding Map. *Benchmark for Science Literacy,* AAAS.
page 71: Graphic by Leigh Coriale Design and Illustration.
page 72: Graphic by Leigh Coriale Design and Illustration.
page 73: Graphic by Leigh Coriale Design and Illustration.
page 83: Graphic by Leigh Coriale Design and Illustration.
page 84: Graphic by Leigh Coriale Design and Illustration.
page 89: Graphic by Leigh Coriale Design and Illustration.
page 102: Graphic by Leigh Coriale Design and Illustration.
page 104: Books, © TSM/Tom Stewart 1995.
page 105: Keyboard, NAP Image Archives.

ORDER CARD
(Customers in North America Only)

Teaching About Evolution and the Nature of Science

Use this card to order additional copies of **TEACHING ABOUT EVOLUTION AND THE NATURE OF SCIENCE** and the books described on the reverse. All orders must be prepaid. Please add $4.00 for shipping and handling for the first copy ordered and $0.50 for each additional copy. If you live in CA, DC, FL, MA, MD, MO, TX, or Canada, add applicable sales tax or GST. Prices apply only in the United States, Canada, and Mexico and are subject to change without notice.

___ I am enclosing a U.S. check or money order.

___ Please charge my VISA/MasterCard/American Express account.

Number: ______________________

Expiration date: ______________________

Signature: ______________________

PLEASE SEND ME:

Qty.	Code	Title	Price
___	TEAEVT	Teaching About Evolution, single copy	$19.95
___	TEAEVO	2-9 copies	$16.50ea*
___	TEAEVO	10+ copies	$13.95ea*
___	SCISTT	National Science Education Standards, single copy	$19.95
___	SCISTB	2-9 copies	$16.50ea*
___	SCISTB	10+ copies	$13.95ea*
___	CHISCI	Every Child a Scientist, single copy	$10.00
___	CHISCI	2-9 copies	$7.00ea*
___	CHISCI	10+ copies	$4.50ea*
___	INTRO	Introducing the NSES (minimum order 50 copies)	$1.50ea*

* No other discounts apply.

Please print.

Name______________________

Address______________________

City ______________ State _____ Zip Code ______________

TOEC

FOUR EASY WAYS TO ORDER

By phone: Call toll-free 1-800-624-6242 or (202) 334-3313 or call your favorite bookstore.

By fax: Copy the order card and fax to (202) 334-2451.

By electronic mail: Order via Internet at http://www.nap.edu/bookstore.

By mail: Return this card with your payment to NATIONAL ACADEMY PRESS, 2101 Constitution Avenue, NW, Lockbox 285, Washington, DC 20055.

Quantity Discounts: 5-24 copies, 15%--25-499 copies, 25%. To be eligible for a discount, all copies must be shipped and billed to one address.

All international customers please contact National Academy Press for export prices and ordering information.

National Science Education Standards

This book offers a coherent vision of what it means to be scientifically literate, describing what all students regardless of background or circumstance should understand and be able to do at different grade levels in various science categories. The book describes the exemplary teaching practices that provide students with experiences that enable them to achieve scientific literacy, criteria for assessing and analyzing students' attainments in science, and the learning opportunities that school science programs afford. In addition, it describes the nature and design of the school and district science program, and the support and resources needed for students to learn science.
ISBN 0-309-05326-9; 1996, 272 pages, 8 x 10.5, index paperbound, single copy, $19.95; 2-9 copies, $16.50 each; 10 or more copies, $13.95 each (no other discounts apply)

Every Child a Scientist

Achieving Scientific Literacy for All

As more schools begin to implement the *National Science Education Standards*, adults who care about the quality of K-12 science education in their communities may want to help their local schools make the transition. This booklet provides guidance to parents and others, explains why high-quality science education is important for all children and young adults, and shows how the quality of school science programs can be measured.
ISBN 0-309-05986-0; 1998, 32 pages, 8.5 x 11, paperbound, single copy, $10.00; 2-9 copies, $7.00 each; 10 or more copies, $4.50 each (no other discounts apply)

Introducing the National Science Education Standards

This booklet provides an overview and background of the vision and principles of the *National Science Education Standards*. Each of the six types of standards is described: content, teaching, assessment, professional development, program, and system. Designed for a general audience, this booklet clarifies what the *Standards* are and responds to typical questions about them.
1997, 16 pages, 8.38 x 9.25, paperbound, $1.50 each, minimum order 50 copies (no other discounts apply)

Use the form on the reverse of this card to order your copies today.

National Science Education Standards

This book offers a coherent vision of what it means to be scientifically literate, describing what all students regardless of background or circumstance should understand and be able to do at different grade levels in various science categories. The book describes the exemplary teaching practices that provide students with experiences that enable them to achieve scientific literacy, criteria for assessing and analyzing students' attainments in science, and the learning opportunities that school science programs afford. In addition, it describes the nature and design of the school and district science program, and the support and resources needed for students to learn science.
ISBN 0-309-05326-9; 1996, 272 pages, 8 x 10.5, index paperbound, single copy, $19.95; 2-9 copies, $16.50 each; 10 or more copies, $13.95 each (no other discounts apply)

Every Child a Scientist

Achieving Scientific Literacy for All

As more schools begin to implement the *National Science Education Standards*, adults who care about the quality of K-12 science education in their communities may want to help their local schools make the transition. This booklet provides guidance to parents and others, explains why high-quality science education is important for all children and young adults, and shows how the quality of school science programs can be measured.
ISBN 0-309-05986-0; 1998, 32 pages, 8.5 x 11, paperbound, single copy, $10.00; 2-9 copies, $7.00 each; 10 or more copies, $4.50 each (no other discounts apply)

Introducing the National Science Education Standards

This booklet provides an overview and background of the vision and principles of the *National Science Education Standards*. Each of the six types of standards is described: content, teaching, assessment, professional development, program, and system. Designed for a general audience, this booklet clarifies what the *Standards* are and responds to typical questions about them.
1997, 16 pages, 8.38 x 9.25, paperbound, $1.50 each, minimum order 50 copies (no other discounts apply)

Use the form on the reverse of this card to order your copies today.